UNRAVELING THE SECRETS OF HUMANITY

A JOURNEY THROUGH TIME, SCIENCE, AND, WONDER

FELIX ATOH

Copyright © 2024 by Felix Atoh

ISBN: 978-1-77883-406-6 (Paperback)

978-1-77883-407-3 (E-book)

BookSide Press
877-741-8091
www.booksidepress.com
orders@booksidepress.com

Unraveling the Secrets of Humanity

Felix Atoh

Contents

Introduction

Setting the Stage

1. The profound curiosity of human beings about their origins and purpose.

Since the dawn of time, humans have gazed at the stars and pondered their place in the cosmos. The relentless quest to understand our origins and purpose has driven us to explore the mysteries of our existence. From ancient myths to modern scientific inquiries, our innate curiosity has shaped the very fabric of our being.

2. The significance of exploring our past, present, and potential future.

As we stand at the crossroads of history, the journey to uncover the secrets of humanity becomes ever more critical. Our past holds the key to understanding the forces that have shaped us, while our present decisions will determine the trajectory of our future. In this momentous journey, we embark on an expedition that transcends time, weaving together the threads of history, science, and wonder.

Objectives of the Book

1. Discovering the hidden mysteries of human history and evolution.

In the first part of this extraordinary voyage, we venture deep into the annals of time to search for our ancestors. From our humble beginnings as primates to the emergence of Homo sapiens, we explore the fascinating puzzle of human evolution. Fossil records and archaeological insights shed light on the ancient paths our ancestors trod, unraveling the enigmatic threads that connect us to our past.

2. Unveiling the scientific breakthroughs that have shaped humanity's progress.

In the second part of our journey, we embark on a scientific odyssey that has revolutionized our understanding of the universe. The Scientific Revolution marked a turning point in human history, leading to remarkable breakthroughs that reshaped our worldview. We uncover the stories of visionary figures and their groundbreaking discoveries that have paved the way for modern science and technology.

3. Inspiring wonder and awe for the remarkable achievements and potential of our species.

Our expedition culminates in a celebration of the wonders of humanity. From the vast expanse of the cosmos to the intricate workings of the human mind, we bask in the awe-inspiring achievements of our species. As we explore the potential of technology, the mysteries of consciousness, and the urgent need to preserve our planet, we are left with a profound sense of wonder at the limitless possibilities that lie ahead.

With each chapter, we seek not only to inform but also to ignite the flames of curiosity within the hearts of our readers. We invite you to join us on this transformative odyssey, to unravel the secrets of humanity and forge a deeper connection with our shared past and future. Let us embark on this journey of discovery, guided by the spirit of exploration and the unyielding thirst for knowledge. As we journey together through time, science, and wonder, may we find inspiration, enlightenment, and a renewed appreciation for the extraordinary tapestry of human existence.

Welcome to "Unraveling the Secrets of Humanity: A Journey through Time, Science, and Wonder."

Part One

The Origins of Humanity

Chapter 1

In Search of Our Ancestors

The quest to trace our roots: from primates to Homo sapiens

In the vast tapestry of time, the origins of humanity remain an enigmatic puzzle. The journey to understand our roots takes us back millions of years, tracing the evolutionary path that has shaped us into the beings we are today. This chapter delves deep into the captivating story of human evolution, exploring the profound connection between modern humans and our ancient ancestors.

UNRAVELING THE PRIMATE ANCESTRY

Our search for our origins begins with our primate cousins. We venture into the lush landscapes where early primates first emerged, observing their adaptations and behaviors that laid the groundwork for our evolutionary journey.

Through genetic analysis and fossil evidence, we piece together the remarkable story of our shared ancestry with primates, revealing the common threads that bind us to our distant relatives.

The Hominin Lineage

Within the realm of early hominins, we uncover the first hints of our unique identity. From Ardipithecus to Australopithecus, we explore the diverse range of hominin species that once roamed the Earth. With each discovery, we gain valuable insights into their anatomy, tools, and social structures, providing essential clues about our ancient past and the factors that drove our evolution.

Homo: The Emergence of Our Genus

Image of an Australopithecus

The advent of the Homo genus marks a pivotal moment in our evolutionary history. We delve into the evidence surrounding Homo habilis, Homo erectus, and other early Homo species, seeking to understand the cognitive leaps that set them apart from their predecessors. Their ability to create tools, control fire, and exhibit complex social behaviors laid the foundation for the emergence of modern humans.

Fossil records and archaeological evidence shed light on human evolution.

The pages of time are etched with evidence of our past, waiting to be unearthed and deciphered. This section dives into the fascinating world of paleontology and archaeology, illuminating the discoveries that have transformed our understanding of human evolution.

The Paleontological Time Capsule

Fossil records are like time capsules, preserving fragments of our ancestors' lives for millennia. We explore the methodologies used by paleontologists to unearth these precious remnants and analyze their significance. From Lucy's remarkable skeleton to the footprints at Laetoli, each discovery offers a glimpse into the lives of our ancient forebears, helping us piece together the puzzle of human evolution.

The Paleontological Time Capsule is a metaphorical concept that refers to the fossil records found in the Earth's geological layers, which act as time capsules preserving fragments of our ancestors' lives over vast periods of time. These fossils are crucial in providing valuable insights into the evolution of life on Earth, including the story of human evolution. Paleontologists, scientists who study ancient life through the examination of fossils, use various methodologies to unearth these precious remnants and analyze their significance.

One of the most remarkable and well-known discoveries in the field of human evolution is the fossilized skeleton of "Lucy." Lucy, officially known as Australopithecus afarensis, was discovered in Ethiopia in 1974. This nearly complete skeleton is over three million years old and represents a key transitional species in human evolution. Lucy's bipedal posture and features provided critical evidence supporting the idea that our ancient ancestors began walking on two legs long before they developed more human-like brain sizes.

Another significant discovery contributing to our under-standing of human evolution is the footprints found at Laetoli in Tanzania. These footprints, dated to around 3.6 million years ago, were preserved in volcanic ash and revealed clear evidence of early hominins walking upright with a human-like stride. The Laetoli footprints offered compelling proof that bipedalism was a defining characteristic of our ancient ancestors.

As readers journey through the pages of the book, they will encounter many other significant discoveries that have enriched our understanding of human evolution. These discoveries may include other fossil finds of early hominins, ancient tools, and artifacts. Each piece of the puzzle uncovered by paleontologists contributes to a deeper discernment of the lives and behaviors of our ancient forebears.

The methodologies used by paleontologists involve careful excavation, detailed analysis, and comparison with existing fossil evidence. Excavation processes often require delicate precision to preserve the fossils intact, as well as accurate recording of the geological context and position of each find. The analysis of fossils includes detailed examinations of their anatomical features, such as teeth, skull structures, and limb bones, to identify the species and understand their evolutionary significance.

By piecing together the information from various fossil discoveries and geological data, scientists can construct a coherent narrative of human evolution. This process allows readers of the book to gain a comprehensive understanding of how early hominins lived, evolved, and adapted to their environments over millions of years.

As the story unfolds, readers will develop an appreciation for the remarkable discoveries that offer glimpses into the distant past, helping to bridge the gap between our ancient ancestors and modern humans. The Paleontological Time Capsule serves as a bridge through time, allowing readers to witness the fascinating journey of human evolution, marvel at the ingenuity of our ancient forebears, and gain a deeper appreciation for the complexity of life's history on Earth.

Archaeological Insights into Ancient Cultures

Beyond fossils, the study of artifacts and ancient settlements brings to life the daily existence of our ancestors. We delve into archaeological sites such as Gobekli Tepe, Olduvai Gorge, and Stonehenge, marveling at the technological achievements and cultural practices of past societies. These insights shed light on the social dynamics, art, and rituals that shaped human communities throughout history.

Theories and controversies surrounding the emergence of modern humans

As we journey further into the depths of human history, we encounter a landscape of diverse theories and controversies that surround the emergence of modern humans. This section presents some of the most debated ideas in the field of anthropology, provoking thought and sparking curiosity.

Out of Africa vs. Multiregionalism

The question of how and where modern humans originated has long been a topic of heated debate. We explore the competing theories of "Out of Africa," which posits that Homo sapiens emerged in Africa and migrated to other parts of the world, and "Multiregionalism," which suggests that modern humans evolved simultaneously in different regions. As we weigh the evidence, we consider the implications of each hypothesis on our understanding of human ancestry.

INTERBREEDING WITH OTHER HOMININS

The genetic analysis of ancient hominin DNA has revealed surprising revelations about the interbreeding between different hominin species, such as Neanderthals and Denisovans. We delve into the fascinating world of ancient genetics, exploring the genetic legacy left behind by our extinct relatives and the impact it has had on the genetic diversity of present-day humans.

As we delve into the depths of time, we find ourselves immersed in the awe-inspiring story of our ancestors. The quest to trace our roots takes us on a journey of discovery, where each fossil, artifact, and scientific revelation adds another brushstroke to the grand mural of human evolution. Join us as we unveil the secrets buried in the layers of the Earth and embark on an unforgettable exploration of our ancient past, seeking to understand the very essence of what it means to be human.

This book "Unraveling the Secrets of Humanity: A Journey through Time, Science, and Wonder" takes a look into the fascinating question of human origins, exploring the two competing theories that have sparked intense debate in the scientific community: "Out of Africa" and "Multiregionalism."

The Out of Africa theory, also known as the recent African origin hypothesis, proposes that Homo sapiens emerged in Africa and subsequently migrated to other parts of the world, gradually replacing other hominin species, such as Neanderthals and Denisovans. This theory suggests that all modern humans today can trace their ancestry back to a single African population that lived around 200,000 years ago. The migration out of Africa occurred in several waves, leading to the peopling of different regions around the globe.

On the other hand, Multiregionalism, or the regional continuity hypothesis, posits that modern humans evolved simultaneously in different regions, maintaining significant gene flow between populations over time. According to this theory, different populations of Homo erectus in various parts of the world evolved independently into Homo sapiens, resulting in a diverse and interconnected network of human populations.

Throughout the book, readers are taken on a journey through the compelling evidence supporting both theories. The evidence in favor of the Out of Africa theory includes genetic studies, which show that non-African populations carry genetic markers that suggest a common African origin. Fossil evidence from Africa also provides support for this hypothesis, as some of the oldest and most primitive Homo sapiens fossils have been discovered on the continent.

In contrast, Multiregionalism proponents point to

archaeological evidence, such as the continuity of certain cultural traits across regions, as well as some genetic studies suggesting gene flow between different populations. They argue that the presence of distinct regional characteristics in modern human populations implies independent evolution in different areas.

As the evidence unfolds, the book carefully considers the implications of each hypothesis on our understanding of human ancestry. The Out of Africa theory highlights the importance of Africa as the cradle of humanity and emphasizes the shared ancestry of all modern humans. It paints a picture of a single, interconnected human family that originated in Africa and embarked on a remarkable journey of exploration and settlement across the globe.

In contrast, Multiregionalism challenges the notion of a singular origin and suggests a more complex and diverse evolutionary process. It recognizes the significance of regional adaptations and interactions between different populations, leading to the rich tapestry of human diversity we see today.

The book acknowledges that the debate between Out of Africa and Multiregionalism is far from settled, and ongoing research continues to refine our understanding of human origins. By exploring both perspectives, "Unraveling the Secrets of Humanity" invites readers to engage with the complexities of our evolutionary history and appreciate the wonder of human diversity and shared ancestry. Ultimately, the journey through time and science in this book leaves readers with a profound appreciation for the mystery and complexity of our species' origin, inviting them to continue exploring the secrets of humanity for themselves.

Chapter 2

Unearthing Ancient Civilizations

The rise of early civilizations: Mesopotamia, Egypt, Indus Valley, and more.

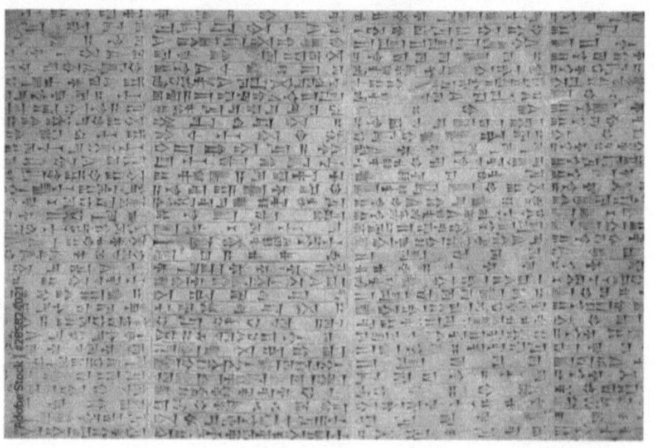

Image of hieroglyphics

In the shadows of time, ancient civilizations emerged as beacons of human ingenuity and resilience, laying the foundation for the complex societies we inhabit today. Among the most remarkable early civilizations were Mesopotamia, Egypt, and the Indus Valley, each with its unique story and contributions to human history.

Mesopotamia: Cradle of Civilization

We journey to the fertile lands between the Tigris and Euphrates rivers, where the civilization of Mesopotamia thrived. From the Sumerians' cuneiform writing to the achievements of the Babylonians and Assyrians, we explore the innovations that transformed Mesopotamia into one of the world's first urban societies. The development of agriculture, trade, and governance in this region set the stage for the trajectory of human history.

Located in the fertile lands between the Tigris and Euphrates rivers, Mesopotamia is often referred to as the cradle of civilization. Around 3500 BCE, the Sumerians settled in this region and developed one of the world's earliest urban societies. They created the world's first known writing system, cuneiform, which enabled the recording of economic transactions, laws, and religious texts. The Sumerians also established city-states and built magnificent ziggurats and temples dedicated to their deities. Mesopotamia's legacy includes significant advancements in mathematics, astronomy, and irrigation, which profoundly influenced later civilizations.

Ancient Egypt: A Gift of the Nile

In the sun-bathed land of Egypt, we uncover the mysteries of one of the most enduring civilizations in history. From the construction of the awe-inspiring pyramids to the development of a sophisticated writing system, the ancient Egyptians left behind a cultural legacy that continues to captivate the world. We delve into their religious beliefs, societal structures, and engineering marvels, revealing the essence of this great civilization.

Ancient Egypt, flourishing along the Nile River around 3000 BCE, is renowned for its awe-inspiring pyramids, monumental architecture, and intricate hieroglyphic writing. The Nile River provided the lifeblood of this civilization, supporting agriculture, trade, and transportation. The Egyptians believed in an afterlife and invested enormous resources in constructing tombs and temples to honor their pharaohs and deities. Their belief in an orderly cosmic system fostered advancements in mathematics and astronomy, vital for their agricultural calendar. The process of mummification demonstrated their early understanding of anatomy and preservation techniques.

INDUS VALLEY CIVILIZATION: LOST IN TIME

The enigmatic Indus Valley Civilization emerges from the mists of antiquity, inviting us to unlock its secrets. Flourishing along the Indus River basin, this advanced society showcased remarkable urban planning, standardized weights, and sophisticated drainage systems. Yet, despite its accomplishments, the Indus Valley Civilization eventually faded into obscurity, leaving behind tantalizing clues of its influence on the region.

Around 2600 BCE, the Indus Valley Civilization flourished in present-day Pakistan and northwestern India. One of the world's earliest urban civilizations, it boasted well-planned cities with advanced drainage systems, indicating a sophisticated level of town planning and public infrastructure. The Indus script remains undeciphered, limiting our understanding of their writing and literature. Their trading activities extended to Mesopotamia and other regions, evidence of a well-connected and prosperous society. The decline of this civilization remains a mystery, adding an air of intrigue to its remarkable story.

Understanding the societal and technological advancements of ancient cultures

Ancient civilizations were cradles of innovation and progress, forging the foundations of human society. In this section, we delve into the societal and technological advancements that shaped their growth and evolution.

These ancient civilizations shared common themes such as monumental architecture, complex societies, written languages, and religious practices. They demonstrated ingenuity in creating agricultural systems, monumental structures, and advancements in science and mathematics.

The stories of these early civilizations serve as a testament to human adaptability and resilience. They faced challenges such as environmental changes, conflicts, and the rise and fall of empires. Yet, their legacies endure in the cultural, technological, and social achievements that laid the groundwork for the modern world.

The study of these civilizations provides valuable insights into the complexity of human societies and the enduring desire for knowledge and understanding. Their achievements and struggles inspire us to appreciate our shared heritage and remind us of the continuity of human progress throughout history. By unveiling and understanding the remarkable stories of these early civilizations, we gain a deeper appreciation for the human journey and the diverse tapestry of our collective past.

Urbanization and Governance

We examine the remarkable transition from small settlements to thriving urban centers in ancient civilizations. The emergence of organized governance, legal systems, and hierarchical structures laid the groundwork for the development of complex societies. As we study their governing principles, we gain valuable insights into the roots of modern political systems.

Image of Egyptian wall art

Urbanization and Governance in Mesopotamia, Egypt, Indus Valley, and other ancient civilizations marked a pivotal transition from small settlements to thriving urban centers. As these societies grew, they developed organized governance, legal systems, and hierarchical structures, which laid the ground work for the development of complex societies. These governing principles and systems served as the roots of modern political systems in many ways.

MESOPOTAMIA

In Mesopotamia, urbanization began around 3500 BCE, with the rise of city-states like Sumer and Akkad. As populations increased, the need for organized governance arose. Each city-state had its own ruler, often a king or priest- king, who claimed divine authority to rule. The king was responsible for maintaining law and order, managing resources, and leading military campaigns. Early legal codes, such as the Code of Hammurabi (c. 1754 BCE), provided a basis for justice and outlined punishments for various offenses. Bureaucratic structures emerged to administer the growing city-states efficiently. Temples played a central role in governance, acting as economic and administrative centers, and supporting the belief in divine authority.

ANCIENT EGYPT

Egypt's urbanization occurred around 3000 BCE, with the emergence of the Nile Valley civilization. The pharaoh was the central figure in Egyptian governance, seen as both a divine ruler and a mediator between gods and humans. The pharaoh exercised absolute authority over the land and its resources. Egypt's bureaucracy, including viziers and scribes, facilitated administrative functions such as taxation, resource management, and record-keeping. The legal system was closely tied to religious beliefs, emphasizing social order and justice. Ma'at, the concept of balance and truth, formed the basis of Egyptian law. Legal proceedings were overseen by local judges and officials.

Indus Valley Civilization

The Indus Valley Civilization, from around 2600 BCE, saw the emergence of organized urban centers like Harappa and Mohenjo-Daro. Little is known about their governance, as the Indus script remains undeciphered. However, the impressive city planning, sanitation systems, and uniformity across settlements suggest a centralized authority. Governance likely involved a ruling elite or council overseeing local affairs. Trade and exchange networks indicate a coordinated system of resource management.

The transition from small settlements to urban centers required effective governance to maintain order, allocate resources, and protect populations. Hierarchical structures emerged to manage social and economic complexities. Kings or rulers held significant power, often legitimizing their authority through religious beliefs. Bureaucracies facilitated administration, record-keeping, and decision-making.

The legal systems of these civilizations were grounded in religious beliefs, with justice seen as essential for maintaining social harmony. The establishment of written laws, like the Code of Hammurabi, showcased early attempts at codifying rules and ensuring consistency in judgments. These legal codes laid the groundwork for the development of modern legal systems.

The principles of governance in these ancient civilizations constitute the roots of modern political systems in several ways. The idea of divine rulership and central authority influencing governance can be seen in various historical and contemporary political systems. Bureaucratic structures and legal codes found their way into later governing systems, contributing to the administrative frameworks of modern governments. Furthermore, the emphasis on justice and social order remains a foundational concept in contemporary legal systems.

Urbanization and governance in ancient civilizations played a crucial role in shaping complex societies and setting the stage for modern political systems. The transition from small settlements to thriving urban centers required organized governance, legal systems, and hierarchical structures to manage the growing complexities of society. The governing principles of these civilizations, rooted in divine rulership, written laws, and social order, continue to influence governance and legal systems worldwide. Studying these ancient political systems provides valuable insights into the evolution of human societies and the enduring impact of their governing principles on the course of history.

Technological Marvels

From agricultural innovations to architectural wonders, ancient civilizations were pioneers of technology in their time. We explore their irrigation methods, astronomical observations, and construction techniques, all of which contributed to their prosperity and cultural achievements. Their inventions continue to inspire modern engineering and technology.

The ancient civilizations of Mesopotamia, Egypt, the Indus Valley, and others were pioneers of technology in their time, leaving behind a legacy of technological marvels that continue to inspire modern engineering and technology. From agricultural innovations to architectural wonders, these civilizations made significant contributions to human progress.

MESOPOTAMIA

In Mesopotamia, the invention of writing (cuneiform) revolutionized communication and record-keeping. This system of wedge-shaped characters impressed on clay tablets enabled the preservation of knowledge, administrative records, literature, and legal codes like the famous Code of Hammurabi. Additionally, the Mesopotamians developed advanced irrigation techniques to harness the waters of the Tigris and Euphrates rivers, enabling them to transform arid lands into fertile agricultural regions. The construction of ziggurats and temple complexes demonstrated their architectural prowess and served as centers of religious and administrative activities.

ANCIENT EGYPT

The ancient Egyptians excelled in architecture, constructing massive pyramids, temples, and monuments that continue to awe the world today. Their engineering skills allowed them to align these structures with astronomical events, showcasing their sophisticated understanding of celestial observations. The invention of papyrus, a form of early paper made from reeds, facilitated the preservation and dissemination of knowledge, literature, and administrative documents. Additionally, the Nile's annual flooding enabled efficient agricultural practices, leading to abundant food production and a stable economy.

INDUS VALLEY CIVILIZATION

The Indus Valley Civilization displayed remarkable urban planning and advanced drainage systems in cities like Harappa and Mohenjo-Daro. Their meticulously planned streets, houses, and public buildings reflected their expertise in city design and sanitation. Archaeological findings suggest that they knew weights and measures, as seen in their standardized brick sizes, indicating a level of technological sophistication.

CHINA (SHANG AND ZHOU DYNASTIES)

Ancient China witnessed significant technological

innovations during the Shang and Zhou Dynasties. The development of bronze metallurgy led to the creation of elaborate bronze vessels, weapons, and ceremonial objects. The use of horse-drawn chariots revolutionized warfare and transportation. Chinese inventors also discovered the technology of silk production, which became a highly prized commodity, contributing to China's economic prosperity and cultural influence.

These technological marvels played a vital role in the prosperity and cultural achievements of these ancient civilizations. Advanced agricultural practices ensured food security, enabling urbanization and cultural growth. Astronomical observations helped in the development of calendars and contributed to the study of celestial phenomena, supporting religious beliefs and cultural practices.

The architectural wonders, including pyramids, ziggurats, and temple complexes, not only served as religious centers but also showcased the architectural ingenuity and engineering capabilities of these civilizations. Their urban planning, drainage systems, and standardized brick sizes displayed an advanced understanding of city design and infrastructure.

Moreover, the inventions of writing systems allowed for the preservation of knowledge, legal systems, literature, and cultural heritage. These inventions served as cornerstones for the growth of education, literature, and administration, influencing the development of subsequent civilizations.

Today, the technological achievements of these ancient civilizations continue to inspire modern engineering and technology. The concept of urban planning and advanced drainage systems has influenced city planning and infrastructure development in modern times. The importance of agricultural innovations in ensuring food security remains relevant as humanity faces the challenges of feeding a growing global population.

In a nutshell, the technological marvels of Mesopotamia, Egypt, the Indus Valley, and other ancient civilizations have left an indelible mark on human history. Their innovations in agriculture, architecture, astronomy, and writing systems contributed to their prosperity and cultural achievements. Moreover, these inventions continue to inspire and influence modern engineering and technology, underscoring the enduring impact of their technological prowess on the course of human progress.

ART, LITERATURE, AND CULTURE

The art, literature, and culture of Mesopotamia, Egypt, the Indus Valley, and other ancient civilizations provide a fascinating glimpse into their beliefs, values, and experiences. Through art and literature, ancient civilizations expressed their beliefs, values, and experiences. We unravel the artistic masterpieces, epic poems, and mythologies that provided a window into their worldviews. Understanding their cultural expressions allows us to glimpse the aspirations and concerns of these ancient societies.

The legacy of ancient civilizations in shaping our present-day societies

Though the echoes of ancient civilizations have long faded, their legacies continue to resonate throughout the corridors of time. In this section, we explore the lasting impact of these ancient societies on the world we live in today.

LANGUAGE AND WRITING

The written word proved to be a transformative legacy of ancient civilizations. We investigate the evolution of writing systems from pictographs to alphabets, revealing the origins of our modern languages and communication methods. The written records left by these civilizations enable us to bridge the gap between the past and the present. The development of writing systems was a transformative achievement of ancient civilizations. From Sumerian cuneiform in Mesopotamia to Egyptian hieroglyphics and the yet-undeciphered script of the Indus Valley, these writing systems allowed for the recording of religious texts, historical events, laws, literature, and administrative records. The decipherment of these scripts has enabled us to reconstruct the histories and beliefs of these ancient societies, bridging the gap between the past and the present. Furthermore, the evolution of these writing systems laid the groundwork for the development of modern languages and communication methods.

Cultural Influence and Traditions

The cultural heritage of ancient civilizations still permeates our global society. We discover how their beliefs, traditions, and artistic expressions have influenced contemporary customs and practices. From religious beliefs to architectural influences, these echoes of the past continue to shape our identities and ways of life. The cultural legacies of these ancient civilizations have left enduring imprints on contemporary customs and practices. Religious beliefs from ancient Egypt and Mesopotamia have influenced modern monotheistic religions, such as Christianity and Islam. The architectural marvels, like the pyramids and ziggurats, continue to inspire modern architecture and engineering. Artistic motifs and styles from these ancient societies have found their way into modern art, design, and fashion. Moreover, their agricultural practices and knowledge of irrigation have shaped modern farming techniques. These echoes of the past contribute to our identities and ways of life, creating a cultural tapestry that weaves together the ancient and the modern.

Lessons from the Past

The rise and fall of ancient civilizations offer valuable lessons and warnings for our own societies. By studying their successes and challenges, we gain insights into the factors that contribute to societal flourishing or decline. The wisdom of the past serves as a guide for navigating the complexities of our present-day world. For example, the collapse of the Indus Valley Civilization, possibly due to environmental factors, serves as a warning about the importance of sustainability and environmental stewardship. The ability of ancient societies to create systems of governance and social organization provides us with a deeper understanding of the complexities of maintaining stable and thriving societies. These lessons from the past act as guides for navigating the complexities of our present-day world, helping us to make informed decisions and shape a better future.

In conclusion, the art, literature, and culture of ancient civilizations like Mesopotamia, Egypt, and the Indus Valley offer a rich tapestry of human history and cultural expression. Their language and writing systems connect us to the past, allowing us to understand their beliefs and experiences. The lasting influence of their cultural heritage continues to shape contemporary customs and practices, and the lessons from their rise and fall serve as valuable insights for our present and future societies. The legacies of these ancient civilizations resonate throughout the corridors of time, leaving an enduring impact on the world we live in today.

As we unearth the wonders of ancient civilizations, we encounter the brilliance of human ingenuity and the resilience of societies that thrived in the face of challenges. Their stories remain etched in the annals of time, inspiring us to learn from the past as we journey forward. Stay with us as we continue our expedition through time, science, and wonder, where each chapter uncovers the hidden marvels of our shared human heritage.

But first, this question:

As we look back at the collapse of the Indus Valley Civilization, does this lesson serve as a call to action against the threat of climate change that looms in the horizon?

Part Two

The Scientific Journey of Humanity

Chapter 3

The Revolution of Science

The Scientific Revolution and its impact on human understanding

In the annals of human history, a profound transformation took place that forever altered our perception of the world and our place in it. This chapter digs deeper into the epochal event known as the Scientific Revolution, a period of intellectual awakening that shattered long-held beliefs and paved the way for a new era of human understanding.

The Scientific Revolution, which emerged as a culmination of human understanding from the era of Mesopotamia, Egypt, the Indus Valley, and other ancient civilizations, had a profound impact on breaking the shackles of dogma and liberating human minds, laying the foundation for modern science.

Image of Digital Revolution

Breaking the Shackles of Dogma

For centuries, prevailing beliefs were heavily influenced by religious and philosophical dogma. The acceptance of established authorities and reliance on ancient texts hindered progress and limited the pursuit of knowledge. The Scientific Revolution challenged this stagnant paradigm by promoting empirical evidence and systematic observation as the basis for understanding the natural world. Visionaries like Nicolaus Copernicus, Galileo Galilei, and Johannes Kepler challenged the geocentric model of the universe, paving the way for the heliocentric model through evidence-based observation. This shift in thinking encouraged scientists to question preconceived notions and embrace skepticism, leading to a transformative liberation of human minds from the constraints of dogma.

The Rise of the Scientific Method

At the heart of the Scientific Revolution was the development of the scientific method—a systematic approach to experimentation, observation, and analysis. The scientific method empowered scientists to formulate hypotheses, design-controlled experiments, and gather empirical evidence. This revolutionary approach allowed for objective testing of ideas and theories, enabling scientists to draw conclusions based on empirical evidence rather than mere speculation. Figures like Francis Bacon and René Descartes played pivotal roles in formalizing the scientific method, emphasizing the importance of rigorous observation and evidence-based reasoning. This shift in methodology marked a turning point in the way knowledge was acquired and validated, ushering in an era of objective inquiry and discovery.

Key figures and breakthroughs that shaped modern science

The Scientific Revolution was not merely an abstract concept, but a journey fueled by the brilliance of visionary minds. In this section, we introduce the key figures whose groundbreaking discoveries transformed the scientific landscape. Before we delve into the work of these key figures, let's take a look at the concept of the Geocentric model.

THE GEOCENTRIC WORLDVIEW

For centuries, the geocentric model, championed by ancient Greek philosophers like Ptolemy, dominated human understanding of the cosmos. This model depicted Earth as the stationary center of the universe, with celestial bodies, including the Sun, Moon, and planets, orbiting around it in complex, epicycle-driven paths. This geocentric perspective was deeply rooted in both religious and philosophical traditions, as it aligned with the Aristotelian view of the universe's organization. All that began to change as more scientific findings became available. Some prominent agents of change include:

1. NICOLAUS COPERNICUS AND THE HELIOCENTRIC MODEL:

Nicolaus Copernicus challenged the geocentric model of the universe by proposing a heliocentric system, where the Earth orbited the Sun. This ground breaking idea laid the groundwork for modern astronomy and shifted humanity's perspective of the cosmos. We delve into his courageous challenge to the geocentric worldview, which placed the Sun at the center of the cosmos and set the stage for the Copernican Revolution.

Nicolaus Copernicus's heliocentric model revolutionized our understanding of the solar system by challenging the prevailing geocentric worldview, which positioned Earth at the center of the cosmos. His courageous challenge to this established belief marked a turning point in humanity's perception of its place in the universe, setting the stage for the Copernican Revolution and reshaping the foundations of modern astronomy. This paradigm shift marked a turning point in humanity's perception of its place in the universe.

In the early 16th century, Nicolaus Copernicus, a Polish

mathematician and astronomer, presented a revolutionary alternative—the heliocentric model. In this model, Copernicus proposed that the Sun, not Earth, was at the center of the solar system. He suggested that Earth and the other planets orbited the Sun in circular orbits, a concept known as the heliocentric hypothesis. This idea challenged the prevailing cosmology and offered a simpler explanation for the observed motions of the planets.

CHALLENGING THE STATUS QUO

Copernicus's heliocentric model was revolutionary for several reasons. Firstly, it provided a more elegant and mathematically coherent explanation for the apparent retrograde motions of the planets, which were difficult to account for in the geocentric model. Secondly, it placed the Sun, a seemingly ordinary star, at the center of the system, shifting the focus from Earth's perceived central importance. This challenge to the traditional geocentric worldview was daring and required courage to present, as it contradicted both religious teachings and established scientific beliefs of the time.

THE COPERNICAN REVOLUTION

While Copernicus's work "De Revolutionibus Orbium Coelestium" (On the Revolutions of the Celestial Spheres) was published in 1543, the heliocentric model faced resistance and skepticism. However, his work laid the foundation for a paradigm shift in astronomical thinking. It sparked further investigation and debate, leading to the

Copernican Revolution, where subsequent astronomers like Johannes Kepler and Galileo Galilei refined and expanded upon Copernicus's ideas.

IMPACT ON HUMANITY'S PERCEPTION

The Copernican Revolution profoundly changed humanity's perception of its place in the universe. It marked a transition from an Earth-centric perspective to a heliocentric one, emphasizing the idea that Earth was not the center of all cosmic motion. This shift in perception challenged traditional religious and philosophical beliefs, prompting a reevaluation of humanity's significance within the vast cosmos. It paved the way for a more scientific approach to understanding celestial phenomena and initiated a new era of astronomical exploration.

In conclusion, Nicolaus Copernicus's heliocentric model ignited a revolution in our understanding of the solar system, challenging the long-held geocentric worldview. His courageous challenge to the prevailing beliefs set the stage for the Copernican Revolution, which reshaped humanity's perception of its place in the universe. The shift from an Earth-centric to a heliocentric perspective marked a transformative turning point in the history of science and laid the groundwork for the modern astronomical understanding that continues to shape our exploration of the cosmos today.

2. Galileo Galilei and the Telescope:

Galileo Galilei's telescopic observations shattered the barriers of the heavens and brought the cosmos closer to humanity, forever transforming our understanding of the universe. His groundbreaking discoveries, such as the moons of Jupiter and the phases of Venus, provided undeniable evidence in support of Copernicus's heliocentric model. Galileo's work marked the birth of modern observational astronomy and propelled humanity to explore the depths of space with newfound curiosity and precision.

In 1609, Galileo constructed his telescope and turned it toward the night sky. His observations revealed a wealth of celestial phenomena that were previously unseen by the naked eye. Among his most significant discoveries were the four largest moons of Jupiter, now known as the Galilean moons: Io, Europa, Ganymede, and Callisto. These observations challenged the prevailing view that all celestial bodies orbited Earth and provided compelling evidence for the heliocentric model proposed by Copernicus.

Phases of Venus

Another groundbreaking discovery made by Galileo was the observation of the phases of Venus. Through his telescope, he observed that Venus exhibited phases similar to those of the Moon, ranging from crescent to gibbous. These observations directly contradicted the geocentric model, which predicted that Venus would only display crescent phases. The phases of Venus observed by Galileo were consistent with the heliocentric model, where Venus orbits the Sun, and its varying illumination from Earth accounts for the observed phases.

BIRTH OF MODERN OBSERVATIONAL ASTRONOMY

Galileo's telescopic observations marked a paradigm shift in our approach to understanding the cosmos. His meticulous recording of celestial observations and his use of experimental techniques laid the foundation for modern observational astronomy. His work emphasized empirical evidence, direct observation, and quantitative measurements—fundamental principles of the scientific method. Galileo's observations not only provided support for the heliocentric model but also demonstrated the power of telescopes as instruments of discovery, opening new avenues for exploring the universe.

IMPACT AND LEGACY

Galileo's discoveries had far-reaching consequences. His work shattered the traditional view that Earth was the center of all cosmic motion, further eroding the geocentric model. However, his observations were met with resistance from the religious and academic establishment, leading to conflicts with authorities. Despite these challenges, Galileo's dedication to empiricism and his commitment to unveiling the truth left an indelible mark on science.

Galileo's observations initiated a new era of astronomical

exploration, prompting scientists to build more advanced telescopes and continue the quest to uncover the mysteries of the universe. His work laid the groundwork for subsequent advancements in astronomy, including Kepler's laws of planetary motion and Newton's law of universal gravitation. The scientific revolution ignited by Galileo's observations propelled humanity toward a deeper understanding of the cosmos and inspired generations of astronomers to explore the depths of space with ever-improving technology and methods.

In summary, Galileo's telescopic observations revolutionized our understanding of the cosmos, providing tangible evidence that supported Copernicus's heliocentric model. His discoveries of Jupiter's moons and the phases of Venus shattered the barriers of the heavens, propelling humanity into the realm of modern observational astronomy. Galileo's work marked a pivotal moment in the history of science, inspiring the exploration of space and shaping our ongoing quest to comprehend the vast universe that surrounds us.

3. The work of Isaac Newton and the Laws of Motion:

Isaac Newton's laws of motion and universal gravitation revolutionized the field of physics. His work demonstrated that the same laws that governed celestial bodies also applied to earthly phenomena, unifying the heavens and the Earth under one set of principles. His laws of motion and universal gravitation represent a revolutionary breakthrough in the field of physics, profoundly altering our understanding of the natural world. His work not only provided a comprehensive framework for explaining the motion of objects on Earth but also demonstrated that the same fundamental principles governed celestial bodies, effectively unifying the heavens and the Earth under a single set of laws.

1. Laws of Motion

Newton's three laws of motion laid the foundation for a systematic understanding of how objects move. The first law, often referred to as the law of inertia, states that an object at rest remains at rest, and an object in motion continues moving at a constant velocity unless acted upon by an external force. The second law establishes a quantitative relationship between force, mass, and acceleration: $F = ma$. The third law, known as the law of action and reaction, states that for every action, there is an equal and opposite reaction.

2. Universal Gravitation

Newton's law of universal gravitation was a monumental revelation that described the attractive force between all objects with mass. The law mathematically quantified the force of gravity between two objects, demonstrating that this force was proportional to the product of their masses and inversely proportional to the square of the distance between them. This law provided a unified explanation for both the falling of objects on Earth and the motion of planets in the heavens.

Unifying the Heavens and the Earth

One of the most remarkable aspects of Newton's work was his realization that the same physical laws applied universally—both on Earth and in the cosmos. By demonstrating that the force of gravity that caused an apple to fall to the ground was the same force that governed the motion of the Moon around Earth and the planets around the Sun, Newton bridged the gap between terrestrial and celestial phenomena.

This unification of the heavens and the Earth under one set of principles had profound implications. It shattered the longstanding belief that the celestial realm operated under separate rules from the terrestrial realm. Newton's laws of motion and universal gravitation provided a unified frame- work that applied to all objects, regardless of their location in the universe. This insight marked a major departure from earlier philosophies that assigned different laws to Earth and the heavens.

Impact and Legacy

Newton's laws of motion and universal gravitation revolutionized physics by introducing a systematic approach to understanding motion, forces, and the interactions between objects. His work laid the groundwork for classical mechanics, providing a comprehensive framework for describing and predicting a wide range of physical phenomena. These laws remained unchallenged for centuries until the advent of relativity and quantum mechanics.

Newton's unification of the heavens and the Earth under a single set of principles also had far-reaching consequences for philosophy and science. It promoted the idea that the natural world was governed by universal laws that could be discovered through empirical observation and mathematical analysis. This approach fundamentally shaped the scientific method and contributed to the Enlightenment's emphasis on reason and understanding.

In conclusion, Isaac Newton's laws of motion and universal gravitation revolutionized physics by providing a comprehensive framework for explaining motion, forces, and the interactions between objects. His work demonstrated that the same principles applied to both terrestrial and celestial phenomena, unifying the heavens and the Earth under a single set of laws. This unification marked a pivotal moment in the history of science, shaping the way we view the natural world and paving the way for the development of modern physics.

How scientific thinking transformed societies and shaped our worldview

The impact of the Scientific Revolution reverberated far beyond laboratories and observatories. It fundamentally transformed societies and reshaped the way humans perceive the world around them. The application of scientific knowledge revolutionized various fields, from medicine and technology to agriculture and industry. It led to increased prosperity, improved living standards, and longer life expectancies. Moreover, scientific thinking encouraged a shift toward secularism and a focus on reason and empirical evidence, challenging traditional religious and philosophical authority. It catalyzed profound changes in societies and fundamentally shifted the way humans perceive the world around them. The scientific mindset fostered curiosity and openness to new ideas, fueling the rapid progress and innovation that characterize modern societies.

1. Advancements in Technology and Industry

Scientific breakthroughs spurred technological innovations that revolutionized industries and daily life. From the steam engine to electricity, the application of scientific knowledge led to significant advancements in transportation, communication, and manufacturing.

2. Enlightenment and Rationality

The Scientific Revolution's emphasis on reason and evidence-based inquiry contributed to the broader Enlightenment movement. This period was marked by a renewed focus on reason, individualism, and skepticism toward authority. Enlightened thinkers like Voltaire, John Locke, and Immanuel Kant championed rationality and advocated for human rights, democracy, and individual freedoms. The rejection of absolute authority and the embrace of human reason laid the intellectual groundwork for the rise of democratic ideals and the questioning of traditional power structures. The impact of the Enlightenment on society was profound, influencing political thought, social reforms, and the shaping of modern democratic institutions.

3. THE EMERGENCE OF MODERNITY

Scientific thinking acted as a catalyst for the emergence of modernity—an era marked by a deep trust in human reason and a rejection of superstition and dogma. This transformation in worldview significantly shaped politics, philosophy, and culture, giving birth to a new era of human progress. This phenomenon remains one of the most transformative periods in human history, reshaping our understanding of the natural world and redefining our relationship with knowledge and truth. The scientific method provided a powerful tool for exploring the universe and unlocking the secrets of humanity. As we journey through the revolutionary landscape of science, we encounter not only the brilliance of key figures and breakthroughs but also the lasting impact of scientific thinking on the fabric of human society and the contours of our collective consciousness. Stick with us as we continue to unravel the mysteries of humanity's journey through time, science, and wonder, where each revelation brings us closer to understanding our place in the grand cosmic tapestry.

In conclusion, the Scientific Revolution emerged as a transformative force, breaking the shackles of dogma and liberating human minds from the constraints of preconceived notions. The development of the scientific method empowered scientists to test hypotheses and draw conclusions based on empirical evidence, revolutionizing the way knowledge was acquired and validated. This revolution in scientific thinking transformed societies, reshaped worldviews, and contributed to the broader Enlightenment movement, ultimately paving the way for modern scientific advancements, democratic ideals, and rational thinking that continue to shape our world today.

Chapter 4

Unveiling the Secrets of the Universe

The wonders of astronomy: from Copernicus to modern space exploration.

Gazing up at the night sky has ignited the curiosity of humankind since time immemorial. This chapter takes us on an extraordinary journey through the realm of astronomy, from the groundbreaking ideas of Copernicus to the cutting-edge achievements of modern space exploration.

The wonders of astronomy encompass an extraordinary journey that stretches from the groundbreaking ideas of Copernicus to the cutting-edge achievements of modern space exploration. This remarkable odyssey has allowed humanity to transcend Earth's boundaries and reach out to touch the stars, uncovering the mysteries of the cosmos.

GROUNDBREAKING IDEAS OF COPERNICUS AND BEYOND

The journey begins with Nicolaus Copernicus, who challenged the geocentric model and proposed the heliocentric model, placing the Sun at the center of the solar system. In the previous chapter, we saw how this revolutionary concept laid the foundation for modern astronomy by shifting our perspective from an Earth-centric to a heliocentric view. His ideas ignited further exploration of the cosmos, fueled by the development of telescopes and observational techniques.

TECHNOLOGICAL ADVANCEMENTS AND SPACE EXPLORATION

Technological advancements propelled humans into the realm of space exploration. The launch of Sputnik, the world's first artificial satellite, by the Soviet Union in 1957 marked the dawn of the space age. The subsequent achievements of Yuri Gagarin, the first human in space, and the iconic Apollo moon missions, including the historic Apollo 11 landing, showcased humanity's ability to venture beyond our planet.

The development of rockets, satellites, and spacecraft equipped with advanced instruments has enabled us to venture beyond Earth's atmosphere. The evolution of materials, propulsion systems, communication technologies, and navigation systems has significantly expanded our reach into the cosmos.

Rockets, initially developed as weapons during World

War II, became the primary means of launching spacecraft into space. The development of more powerful and efficient rocket engines, such as the Saturn V used in the Apollo missions, allowed for heavier payloads and longer journeys. Innovations in materials science led to the creation of heat-resistant materials for reentry and protective shielding for spacecraft.

Advances in communication technology enabled real-time communication between mission control and spacecraft, ensuring the safety and success of missions. Navigation systems, including GPS, provided accurate positioning and trajectory calculations for spacecraft, enabling precise maneuvers and rendezvous with other celestial bodies.

SPUTNIK AND THE DAWN OF THE SPACE AGE

On October 4, 1957, the Soviet Union launched Sputnik 1, the world's first artificial satellite, into orbit around Earth. This event marked the dawn of the space age and initiated a new era of exploration beyond our planet. Sputnik's launch was a significant achievement in itself, demonstrating the feasibility of launching objects into orbit and paving the way for future space missions.

Sputnik's radio signals were heard around the world, symbolizing the Soviet Union's technological prowess and igniting a space race between the United States and the Soviet Union. The launch of Sputnik not only had scientific and technological implications but also had political and geopolitical ramifications, as it intensified the Cold War rivalry between the two superpowers.

YURI GAGARIN, APOLLO MOON MISSIONS, AND HUMANITY'S VENTURE BEYOND EARTH

On April 12, 1961, Yuri Gagarin, a Soviet cosmonaut, became the first human to travel into space aboard the Vostok 1 spacecraft. Gagarin's historic flight marked a pivotal moment in human history, showcasing our ability to send a person beyond the confines of our planet. His journey orbited Earth, lasting just 108 minutes, but it opened the door to crewed space exploration. This milestone inspired a new era of exploration and ignited the dreams of millions.

The Apollo moon missions, led by NASA, represented the pinnacle of space exploration achievements. The Apollo program's crowning glory was the Apollo 11 mission, during which astronauts Neil Armstrong and Edwin "Buzz" Aldrin became the first humans to set foot on the lunar surface on July 20, 1969. Armstrong's famous words, "That's one small step for [a] man, one giant leap for mankind," echoed the triumph of human determination and ingenuity.

The success of the Apollo 11 landing showcased humanity's ability to overcome immense challenges and achieve audacious goals. It represented a testament to technological innovation, meticulous planning, and international collaboration. The moon landings demonstrated that humans could not only venture beyond Earth but also explore and inhabit other celestial bodies, expanding our horizons and opening the door to further exploration of our solar system and beyond. These successful moon landings not only expanded our understanding of the universe but also inspired generations to dream of exploring the cosmos.

Technological advancements clearly propelled humans into the realm of space exploration, allowing us to overcome the limitations of our planet. The launch of Sputnik marked the beginning of the space age, while Yuri Gagarin's flight and the Apollo moon missions demonstrated humanity's ability to venture beyond Earth's boundaries and reach out to touch the stars. These achievements showcased the remarkable fusion of science, technology, and human endeavor that has expanded our understanding of the universe and paved the way for future exploration.

MODERN SPACE EXPLORATION

With technological advancements, humans reached out to touch the stars. We embark on the history of space exploration, from the first satellite, Sputnik, to the iconic Apollo moon missions. The relentless pursuit of knowledge beyond our planet has led to unprecedented achievements, from exploring distant planets to setting foot on the lunar surface. Modern space exploration continues to push the boundaries of human understanding and inspire the next generation of explorers. Fueled by technological advancements; it has allowed humanity to reach out and touch the stars in once unimaginable ways. This remarkable journey through the cosmos has led to unprecedented achievements, from the launch of the first satellite, Sputnik, to the iconic Apollo moon missions and beyond.

UNPRECEDENTED ACHIEVEMENTS AND ONGOING EXPLORATION

Modern space exploration has indeed witnessed remarkable achievements beyond the historic moon landings. Robotic spacecraft missions like Voyager, New Horizons, and Mars rovers have ventured into the depths of space, unveiling unprecedented insights into distant planets, moons, and celestial bodies. These missions have expanded human knowledge of the solar system's composition, history, and potential for hosting life.

Modern space exploration, driven by technological advancements, has allowed humanity to transcend the boundaries of Earth and touch the stars. From the launch of Sputnik to the Apollo moon missions and beyond, our relentless pursuit of knowledge has led to unprecedented achievements and expanded our understanding of the universe. As we continue to explore and push the boundaries of space, we inspire future generations to explore, innovate, and dream of the limitless possibilities that lie beyond our home planet.

VOYAGER MISSIONS

Launched in 1977, the Voyager spacecraft embarked on an ambitious journey to explore the outer planets and beyond. Voyager 1 and Voyager 2 conducted flybys of Jupiter, Saturn, Uranus, and Neptune, providing detailed images and data about these distant worlds. These missions revealed the intricate details of the gas giants' atmospheres, rings, and moons. Voyager 1, in particular, ventured beyond the solar system's heliosphere, entering interstellar space and continuing to transmit valuable data back to Earth. These missions transformed our understanding of the outer planets and their complex systems.

New Horizons Mission

Launched in 2006, the New Horizons spacecraft completed a historic flyby of Pluto in 2015, offering humanity its first close-up view of the distant dwarf planet. The mission provided high-resolution images and data that unveiled Pluto's diverse landscape, including its icy plains, towering mountains, and unique atmosphere. New Horizons also ventured deeper into the Kuiper Belt, studying a region of the solar system that contains primitive bodies dating back to its formation. This mission revolutionized our understanding of Pluto and the outer reaches of the solar system.

Mars Rovers

NASA's Mars rovers, including Spirit, Opportunity, and Curiosity, have significantly expanded our knowledge of the Red Planet. These rovers have explored the Martian surface, analyzing rocks, soils, and geological features to understand the planet's past environmental conditions.

Opportunity, for example, discovered evidence of past liquid water on Mars, suggesting that the planet might have supported conditions conducive to life in its distant past. Curiosity, a more advanced rover, has been exploring the Gale Crater, analyzing its geology and searching for signs of habitability. These rovers have transformed Mars from a distant, mysterious world to a place where we can envision the possibilities of past or present life.

Expanding Knowledge and Potential for Life

The insights gained from these missions have expanded our understanding of the solar system's composition, history, and potential for hosting life. By studying the geology, atmospheres, and surface conditions of distant planets and moons, scientists have learned about the processes that shaped these bodies over billions of years. The presence of water ice on several celestial bodies, such as Jupiter's moon Europa and Saturn's moon Enceladus, suggests the potential for subsurface oceans and raises tantalizing questions about the possibility of extraterrestrial life.

Additionally, these missions have helped scientists unravel the history of our solar system, shedding light on its formation and evolution. By analyzing the chemical compositions of distant worlds, researchers can infer the conditions that existed during the early stages of the solar system's development.

Yes, modern space exploration has gone beyond the moon landings, utilizing robotic spacecraft to venture into the depths of space and explore distant planets, moons, and celestial bodies. Missions like Voyager, New Horizons, and Mars rovers have expanded our knowledge of the solar system's composition, history, and potential for hosting life. These missions have transformed our understanding of the cosmos, providing invaluable insights into the diverse worlds that inhabit our solar system and inspiring us to continue our quest for knowledge beyond Earth.

The International Space Station (ISS)

The International Space Station serves as a symbol of international cooperation in space exploration. It offers a platform for scientific research, technology development, and testing the effects of long-duration space travel on the human body. The ISS serves as a steppingstone for future missions to deep space, including Mars.

PUSHING BOUNDARIES AND INSPIRING THE FUTURE

Modern space exploration continues to push the boundaries of human understanding. Advancements in propulsion technology, artificial intelligence, and materials science are enabling ambitious plans for crewed missions to Mars and beyond. The pursuit of knowledge beyond our planet not only satisfies our innate curiosity but also drives technological innovations that benefit life on Earth. Rovers like Curiosity and Perseverance are actively exploring Mars, seeking signs of past or present life.

Moreover, space exploration inspires the next generation

of explorers, scientists, engineers, and dreamers. It showcases the power of collaboration, human potential, and the pursuit of audacious goals. As we venture further into space, we continue to unravel the mysteries of the cosmos, shaping our understanding of our place in the universe and inspiring us to dream big and reach for the stars.

Modern space exploration, driven by technological advancements, has allowed humanity to transcend the boundaries of Earth and touch the stars. From the launch of Sputnik to the Apollo moon missions and beyond, our relentless pursuit of knowledge has led to unprecedented achievements and expanded our understanding of the universe. As we continue to explore and push the boundaries of space, we inspire future generations to explore, innovate, and dream of the limitless possibilities that lie beyond our home planet.

The wonders of astronomy encompass an extraordinary journey that stretches from the groundbreaking ideas of Copernicus to the cutting-edge achievements of modern space exploration. This remarkable odyssey has allowed humanity to transcend Earth's boundaries and reach out to touch the stars, uncovering the mysteries of the cosmos.

GROUNDBREAKING IDEAS OF COPERNICUS AND BEYOND

The journey begins with Nicolaus Copernicus, who challenged the geocentric model and proposed the heliocentric model, placing the Sun at the center of the solar system. This revolutionary concept laid the foundation for modern astronomy by shifting our perspective from an Earthcentric to a heliocentric view. His ideas ignited further exploration of the cosmos, fueled by the development of telescopes and observational techniques.

TECHNOLOGICAL ADVANCEMENTS AND SPACE EXPLORATION

Technological advancements propelled humans into the realm of space exploration. The launch of Sputnik, the world's first artificial satellite, by the Soviet Union in 1957 marked the dawn of the space age. The subsequent achievements of Yuri Gagarin, the first human in space, and the iconic Apollo moon missions, including the historic Apollo 11 landing, showcased humanity's ability to venture beyond our planet.

UNVEILING THE COSMOS

Space exploration has unveiled the wonders of our solar system and beyond. Robotic missions have provided closeup views of planets like Mars, revealing ancient river valleys and polar ice caps. Voyager spacecraft have journeyed to the outer reaches of the solar system, capturing images of gas giants and their moons. The Hubble Space Telescope has offered breathtaking views of distant galaxies, nebulae, and other celestial phenomena, expanding our understanding of the universe's vastness and complexity.

Pushing Boundaries and Inspiring Generations

Modern space exploration continues to push the boundaries of human understanding. Rovers like Curiosity and Perseverance are actively exploring Mars, seeking signs of past or present life. The search for exoplanets in habitable zones hints at the possibility of extraterrestrial life. The International Space Station serves as a laboratory for scientific research, technological development, and international cooperation.

Inspiration for the Future

Space exploration inspires the next generation of explorers, scientists, and dreamers. It demonstrates the limitless potential of human ingenuity, the power of international collaboration, and the spirit of discovery. Advances in space technology, such as reusable rockets and plans for crewed missions to Mars, point toward a future where humanity may become a multi-planetary species. The pursuit of knowledge beyond our planet enriches our understanding of Earth, our place in the cosmos, and the interconnectedness of all life.

In conclusion, the wonders of astronomy encompass an awe-inspiring journey from Copernicus's revolutionary ideas to the modern achievements of space exploration. Technological advancements have allowed humans to venture into space, touch the stars, and unveil the cosmos's mysteries. Space exploration pushes the boundaries of human understanding, inspiring generations to reach for the stars and explore the frontiers of knowledge. The relentless pursuit of knowledge beyond our planet continues to shape our understanding of the universe and holds the promise of an exciting and ever-expanding future.

The quest to understand the cosmos, blackholes, dark matter, and beyond

As we peer deeper into the cosmos, we encounter profound mysteries that challenge our current understanding of the universe. In this section, we embark on a cosmic voyage to unravel the secrets that lie beyond our reach.

Image of a black hole

1. THE MYSTERIES OF BLACK HOLES

Black holes, the cosmic enigmas formed from collapsed stars, have captivated both scientists and the public alike. We delve into the nature of black holes, their formation, and their gravitational influence on surrounding matter.

Exploring the paradox of the event horizon and the groundbreaking discoveries of gravitational waves, we witness the fascinating dance between black holes and the fabric of spacetime.

2. DARK MATTER

The Invisible Cosmic Web: The majority of the universe remains elusive, shrouded in darkness. We explore the concept of dark matter, an invisible and mysterious substance that vastly outweighs ordinary matter. Its gravitational effects shape the cosmic web, weaving galaxies together in an intricate tapestry. Although we cannot observe dark matter directly, its presence profoundly impacts the universe's evolution and structure.

Black holes are some of the most enigmatic and awe-inspiring objects in the universe, challenging our current understanding of space, time, and gravity. These regions of space possess such immense gravitational pull that nothing, not even light, can escape from them. While our understanding of black holes has advanced significantly, several profound mysteries persist:

- Singularity and Event Horizon: The very core of a black hole is believed to be a singularity—a point of infinite density where the laws of physics breakdown. Surrounding the singularity is the event horizon, a boundary beyond which nothing can escape. The nature of the singularity and what happens beyond the event horizon remains one of the biggest mysteries.

- Information Paradox: According to our current understanding of physics, information cannot be destroyed. However, black holes seem to violate this principle. The concept of Hawking radiation suggests that black holes emit radiation and lose mass over time, eventually evaporating completely. This raises questions about what happens to the information that falls into a black hole and whether it can be preserved.

- Black Hole Mergers: The detection of gravitational waves from blackhole mergers has provided valuable insights into their existence. However, the process of black hole formation, growth, and mergers is still not fully understood.

How do black holes of different sizes merge, and what are the effects of such mergers on the fabric of spacetime?

Black hole mergers are some of the most cataclysmic and fascinating events in the universe. When black holes of different sizes come together, they create powerful gravitational interactions that reshape the fabric of spacetime and release immense amounts of energy in the form of gravitational waves. Here's an overview of how black holes merge and the effects of these mergers on spacetime:

Process of Black Hole Mergers:

1. Approach and Orbital Decay: When two black holes

are in close proximity, close, their immense gravitational attraction causes them to enter into an orbital dance. Over time, this dance causes their orbits to decay, bringing them closer to each other.

2. Emission of Gravitational Waves: As the black holes orbit each other, they emit gravitational waves—ripples in spacetime caused by the acceleration of massive objects. These gravitational waves carry away energy and angular momentum, causing the black holes to spiral closer together.

3. Coalescence and Final Merge: As the black holes continue to lose energy through gravitational wave emission, their orbits become increasingly tight. Eventually, they reach a point where they merge into a single, more massive black hole. This coalescence results in a violent release of gravitational energy.

Effects on Spacetime:

The merger of black holes has profound effects on the fabric of spacetime:

1. Gravitational Waves: The most immediate effect is the production of gravitational waves. These waves radiate outward in all directions, carrying information about the merging of blackholes' masses, spins, and orbits. Gravitational wave detectors like LIGO (Laser Interferometer Gravitational-Wave Observatory) and Virgo are capable of detecting these waves when they

reach Earth.

2. Curvature of Spacetime: As the black holes merge, they create intense curvature in spacetime around them. This curvature propagates outward with the gravitational waves, temporarily distorting the geometry of space itself.

3. Energy Release: The merger of black holes releases an enormous amount of energy in the form of gravitational waves. This energy is equivalent to several times the mass of our Sun, converted into gravitational radiation.

4. Ringdown Phase: After the merger, the resulting single black hole undergoes a phase called the "ringdown." During this phase, the black hole settles into a stable shape, emitting gravitational waves that gradually dampen over time.

The effects of a black hole merger are an intricate interplay between spacetime curvature, energy release, and the propagation of gravitational waves. These mergers provide a unique opportunity to test Albert Einstein's general theory of relativity and provide insights into the properties of blackholes, such as their masses, spins, and even the nature of the event horizon.

The study of black hole mergers is a fascinating field of research that combines theoretical modeling, computer simulations, and cutting-edge observational techniques. By observing and analyzing the gravitational waves emitted during these events, scientists can unlock new insights into the nature of black holes, the behavior of gravity, and the fundamental structure of the universe.

DARK MATTER (THE INVISIBLE COSMIC WEB)

Dark matter is an invisible form of matter that does not emit or interact with electromagnetic radiation, making it exceptionally difficult to detect. Its presence is inferred from its gravitational effects on visible matter. While dark matter is believed to make up a significant portion of the universe's mass, its true nature remains elusive:

- Particle Nature: Despite its prevalence, the nature of dark matter particles is unknown. Scientists have proposed various hypothetical particles, such as Weakly Interacting Massive Particles (WIMPs), as candidates for dark matter. However, these particles have not yet been observed directly, leaving the true identity of dark matter unresolved.

- Distribution and Structure: Dark matter is thought to form a vast cosmic web that connects galaxies and galaxy clusters. This web guides the formation of galaxies and the large-scale structure of the universe. Mapping the distribution and understanding of the interactions within this web is a complex challenge that requires advanced simulations and observational techniques.

- Interactions with Ordinary Matter: While dark matter does not interact electromagnetically, it may interact weakly with ordinary matter. Efforts are ongoing to directly detect such interactions using sophisticated detectors deep underground. However, no conclusive evidence of dark matter interactions with ordinary matter has been obtained so far.

The mysteries of black holes and dark matter stand as profound challenges to our current understanding of the universe. These enigmatic phenomena continue to captivate scientists and inspire research into the fundamental nature of space, time, and matter. As technology advances and our observational capabilities improve, we hope to unravel these mysteries and gain deeper insights into the nature of the cosmos.

Beyond the Observable Universe

The observable universe is but a fraction of the vast cosmos. We contemplate the philosophical implications of the universe's apparent limit and the question of what lies beyond. We grapple with our limitations as observers, pondering the mysteries of the cosmic horizon and the potential existence of multiverses, parallel dimensions, and other cosmic realms.

The philosophical implications of our place in the vast universe

In the cosmic expanse, human beings are but a speck of dust. This section delves into the profound philosophical questions raised by our understanding of the universe and our place within it.

THE HUMAN PERSPECTIVE

As we explore the cosmos, we reflect on our significance and insignificance in the grand scheme of things. The vastness of the universe can evoke feelings of awe and humility, prompting us to contemplate our place as conscious beings in an ever-expanding cosmos.

LIFE BEYOND EARTH

The revelation of exoplanets and the potential for life beyond our planet ignited speculation about extraterrestrial life. We ponder the philosophical implications of the existence of intelligent life elsewhere in the universe and the profound impact such a discovery could have on humanity's self-perception.

THE SEARCH FOR MEANING

The quest to understand the universe inevitably leads to introspection about the nature of existence and our purpose in life. We explore how the scientific endeavor and the mysteries of the cosmos intersect with philosophical inquiries about the meaning of life and our place in the cosmic drama.

As we unveil the secrets of the universe, we find ourselves at the crossroads of wonder and inquiry. The mysteries that lie beyond our world beckon us to explore and seek knowledge, urging us to grapple with profound questions about our existence and the nature of reality. Come with us as we journey deeper into the cosmos, where each discovery opens new frontiers of understanding and fuels our quest for cosmic wisdom and enlightenment. Beyond the observable universe lies a realm of cosmic mysteries and vast distances that challenge our understanding of the universe and our place within it. This reality gives rise to profound philosophical implications that have intrigued thinkers for centuries. Some of these implications include:

COSMIC INSIGNIFICANCE AND SIGNIFICANCE

The sheer vastness of the universe can evoke feelings of insignificance in human existence. The realization that Earth is just a tiny speck in a seemingly endless expanse of space can lead to questions about the significance of human life and achievements. On the other hand, this perspective can also inspire awe and wonder, emphasizing the preciousness of our existence in the face of cosmic grandeur.

Contemplating the Vast Universe: Evoking Insignificance and Awe

1. THE VAST UNIVERSE AND INSIGNIFICANCE: The sheer vastness of the universe, with its billions of galaxies, each containing billions of stars and planets, can evoke feelings of profound insignificance in human existence. As we peer into the night sky and consider the enormity of cosmic distances, a sense of humility emerges. The realization that our entire planet is a mere speck in the cosmos can leave us questioning our place and purpose

in the universe.

2. Eᴀʀᴛʜ's Tɪɴʏ Pʀᴇsᴇɴᴄᴇ: Significance and Reflection: Considering Earth's position as a small dot amidst an immeasurable expanse of space can lead to deep introspection. The question arises: Do our individual lives and achievements hold any true significance in the face of such cosmic grandeur? The universe's scale prompts philosophical inquiries about human identity, values, and the ultimate meaning of existence. As we grapple with these questions, we may confront existential angst and a search for purpose.

3. Aᴡᴇ ᴀɴᴅ ᴛʜᴇ Pʀᴇᴄɪᴏᴜsɴᴇss ᴏꜰ Exɪsᴛᴇɴᴄᴇ: However, the perspective of Earth's insignificance in the cosmos can also inspire awe and emphasize the preciousness of our existence. Rather than diminishing our significance, it highlights the rarity of life in the universe. Earth's delicate balance, capable of supporting diverse forms of life, becomes a testament to the intricate interplay of cosmic forces that allow life to flourish.

Biological Uniqueness: Recognizing Earth as a haven for life reinforces the idea that life is a remarkable phenomenon in the vastness of space. Our existence becomes a testament to the tenacity of life's emergence and evolution.

Collective Endeavors: Despite our apparent smallness, human achievements become collective accomplishments that transcend individual insignificance. Our collective pursuit of knowledge, exploration, and innovation stands as a testament to human curiosity and determination.

<u>Shared Humanity</u>: This perspective can foster a sense of unity among humans, emphasizing our commonality in the face of cosmic magnificence. We become part of a grand narrative that spans cultures, borders, and time.

These perspectives also carry the potential to inspire awe, reinforcing the preciousness of life and the collective endeavors of humanity. As we navigate these thoughts, we can find a balance between humility and empowerment, drawing strength from our shared journey as inhabitants of a vast and wondrous cosmos.

Top of Form

HUMILITY AND PERSPECTIVE

Contemplating the expansiveness of the universe can foster humility. Recognizing our minuscule place within the cosmos encourages us to transcend personal concerns and embrace a broader perspective. This perspective can encourage unity and cooperation among humanity as we grapple with shared challenges and aspirations.

ACHIEVING HUMILITY AND A BROADER PERSPECTIVE IN THE COSMOS

1. Foster Humility:

Education and Awareness: Promote education about the

vastness of the universe, imparting an understanding of our relative insignificance. Learning about the cosmos humbles us by emphasizing the enormity beyond our immediate surroundings.

Reflective Practices: Encourage practices like stargazing, contemplating the night sky, or engaging with space documentaries. These experiences evoke awe and instill a sense of humility by revealing the grandeur of the universe.

Comparative Thinking: Compare Earth's size to celestial bodies like stars and galaxies. Such comparisons magnify our smallness, fostering humility and shifting our perspective.

2. Embrace Our Place in the Cosmos:

Cosmic Perspective Education: Incorporate the cosmic perspective into curricula, teaching students about their place in the universe from an early age. This perspective can help them transcend personal concerns and appreciate their role in a larger cosmic narrative.

Astronomy and Space Exploration: Promote astronomy clubs, space museums, and engagement with space agencies. Participation in space-related activities fosters a sense of connection to the universe and a desire to contribute to its exploration.

3. Encouraging Unity and Cooperation:

<u>Shared Exploration</u>: Collaborative efforts in space exploration, like the International Space Station, exemplify global cooperation. Highlight such achievements to showcase how humanity can work together to overcome challenges.

<u>Common Cosmic Heritage</u>: Emphasize that all humans share the same cosmic origins, drawing from the same star stuff that constitutes the universe. This perspective can encourage unity by highlighting our shared cosmic heritage.

<u>Challenges and Solutions</u>: Present shared challenges like climate change, pandemics, and resource scarcity as global issues that require collective solutions. Our unity as a species becomes crucial in addressing these challenges.

<u>Community Engagement</u>: Public Lectures and Workshops: Organize public talks and workshops on astronomy, cosmology, and the universe's vastness. Engaging in discussions can promote humility and a broader perspective.

<u>Space-themed Art and Culture</u>: Use art, literature, and cultural events to convey the cosmic perspective. These mediums can evoke awe and encourage contemplation about our place in the cosmos.

<u>Online Astronomy Platforms</u>: Utilize social media and online platforms to share images, videos, and articles about the universe. These platforms provide easy access to cosmic knowledge and inspire a sense of wonder.

<u>Virtual Space Tours</u>: Develop virtual tours that take people on simulated journeys through the cosmos. This technology can bring the cosmic perspective directly to individuals' screens.

In conclusion, achieving humility, embracing our place in the cosmos, and promoting unity can be fostered through education, community engagement, technology, and cultural initiatives. By imparting the cosmic perspective, we can shift our focus from individual concerns to shared aspirations and challenges, encouraging collaboration, unity, and a sense of wonder about the universe we all call home.

THE SEARCH FOR MEANING: SEEKING MEANING IN THE VASTNESS OF THE UNIVERSE

The vastness of the universe prompts questions about the meaning of human existence. Philosophical inquiries into purpose, destiny, and the nature of reality gain deeper significance when placed in the context of an incomprehensibly large cosmos. People may seek meaning in different ways, such as through scientific exploration, spiritual beliefs, or philosophical contemplation. Each approach offers unique perspectives on our place in the cosmos and the purpose of our existence.

1. Scientific Exploration:

Understanding Origins: Scientific endeavors seek to unravel the mysteries of the universe's origin, evolution, and the emergence of life. This pursuit of knowledge offers a sense of purpose by uncovering the fundamental processes that shaped our existence.

Cosmic Connection: Scientific exploration connects humanity to the universe's intricate mechanisms and cosmic events. Observing celestial phenomena provides insights into our place within the grand cosmic narrative.

2. Spiritual Beliefs:

Transcendence: Many spiritual traditions perceive the universe as a manifestation of divine creation. People find meaning by viewing themselves as part of a larger, purposeful plan orchestrated by a higher power.

Connectedness: Spiritual beliefs often emphasize interconnectedness with all living beings and the universe. This interconnectedness can infuse individuals' lives with meaning and a sense of belonging.

3. Philosophical Contemplation:

Existential Inquiry: Philosophers explore questions about the nature of reality, the purpose of life, and human consciousness. This contemplation prompts individuals to confront their existence's deeper aspects and seek meaning in the face of the universe's vastness.

Ethical Reflection: Philosophical inquiry encourages individuals to reflect on their values, ethics, and contributions to the world. The quest for a meaningful life often involves considering how one's actions impact others and the broader cosmos.

4. Artistic Expression:

Creative Interpretation: Artists interpret the universe's grandeur through various forms of expression, from visual arts to literature and music. These interpretations can evoke emotions, inspire awe, and prompt introspection about existence.

5. Community and Connection:

Shared Exploration: People find meaning by connecting with others who share a curiosity about the universe. Joining astronomy clubs, participating in discussions, and attending lectures create a sense of community around cosmic exploration.

Interpersonal Relationships: Relationships with loved ones and the desire to leave a positive impact on the world provide a sense of purpose and meaning, despite the universe's vastness.

6. Environmental Stewardship:

Guardianship of Earth: Viewing Earth as a precious home in the vastness of space inspires efforts to protect the planet. Environmental consciousness and sustainability become sources of meaning.

7. Search for Personal Fulfillment:

Self-Discovery: Individuals seek meaning by exploring their passions, talents, and personal growth. The universe's vastness can inspire them to embrace opportunities and strive for personal fulfillment.

The search for meaning in the vastness of the universe is a deeply personal and multifaceted journey. People navigate this quest through scientific exploration, spiritual beliefs, philosophical contemplation, artistic expression, community connection, and personal fulfillment. Each approach provides a unique lens through which individuals find purpose, connection and a deeper understanding of their place in the cosmos.

THE LIMITS OF KNOWLEDGE

The universe's vastness implies the limits of human knowledge. Our current understanding of the cosmos is constrained by what we can observe and measure. The unobservable regions challenge our ability to fully comprehend the nature of the universe. This limitation highlights the ongoing journey of discovery and the perpetual quest for knowledge.

Does the Universe's Vastness Imply the Limits of Human Knowledge?

The vastness of the universe does not necessarily imply the absolute limits of human knowledge. While the universe's enormity can be overwhelming, it also serves as an incentive for exploration and discovery. Human knowledge has expanded significantly over the centuries, driven by curiosity and advancements in science and technology.

Example: The Hubble Space Telescope has revolutionized our understanding of the cosmos. Its images of distant galaxies and celestial phenomena have allowed scientists to peer billions of years into the past, revealing new insights into cosmic evolution.

Is Our Current Understanding of the Cosmos Truly Constrained by Observation and Measurement?

While our understanding of the cosmos is largely built on observation and measurement, it's important to recognize that not all aspects of the universe are directly observable. The nature of dark matter, dark energy, and blackholes, for instance, pose challenges beyond direct observation.

Example: Dark matter, which makes up about 27% of the universe, doesn't emit light and thus can't be seen directly. However, its presence is inferred through its gravitational effects on visible matter, leading to the formulation of new theories about its composition.

Overcoming the Limitations of Unobservable Regions: The Ongoing Journey of Discovery

TECHNOLOGICAL ADVANCEMENTS

Telescopes and Instruments: Advanced telescopes, like the James Webb Space Telescope (upcoming), will peer deeper into space, capturing more distant and faint objects.

Particle Accelerators: These machines allow researchers to simulate conditions in the early universe and study particle interactions, contributing to our understanding of fundamental physics.

THEORETICAL MODELS

Simulation and Modeling: Scientists use supercomputers to simulate complex cosmic processes, allowing them to study phenomena like galaxy formation and black hole mergers.

Predictive Theories: Theoretical frameworks like string theory and loop quantum gravity propose ways to reconcile general relativity and quantum mechanics, potentially explaining unobservable phenomena.

INTERDISCIPLINARY COLLABORATION

Astrophysics and Particle Physics: Collaborations between these fields enhance our understanding of cosmic phenomena that involve both macroscopic and microscopic scales.

Philosophy and Science: Philosophical inquiry helps conceptualize the limits of human knowledge and the nature of unobservable phenomena.

INNOVATION AND DISCOVERY

Gravitational Waves: The detection of gravitational waves from black hole mergers offers a new way to explore the universe's most extreme events.

Exoplanet Research: Unveiling the Universe's Hidden Worlds: Discovering exoplanets in habitable zones advances our understanding of potential life beyond Earth.

Exoplanet research is a groundbreaking field of astronomy that focuses on the study of exoplanets, also known as extrasolar planets. Exoplanets are planets that orbit stars outside our solar system, and their discovery has transformed our understanding of planetary systems, the potential for extraterrestrial life, and the vastness of the universe. This field involves various methods of detection, characterization, and exploration, offering insights into the diversity of planetary systems beyond our own.

METHODS OF DETECTION

Transit Method: When an exoplanet passes in front of its host star, it causes a temporary decrease in the star's brightness. Observing these periodic dimming events allows astronomers to infer the exoplanet's size and orbit.

Radial Velocity Method: Exoplanets induce small gravitational wobbles on their host stars, causing a periodic shift in the star's spectral lines. By analyzing these shifts, scientists can determine the planet's mass and orbit.

Direct Imaging: Using advanced telescopes, researchers capture images of exoplanets by blocking out the light from their host stars. This method provides insights into the planet's atmospheric composition and characteristics.

Microlensing: Gravitational lensing occurs when a massive object, like a star, bends and magnifies light from a background source. If a planet passes in front of the lensing star, it creates a detectable signal.

CHARACTERIZATION AND EXPLORATION

Atmospheric Composition: Spectroscopic analysis of an exoplanet's atmosphere reveals its chemical composition. Identifying molecules like water vapor, carbon dioxide, and methane can provide clues about their potential habitability.

Habitability Assessment: Researchers study the distance

between an exoplanet and its host star's habitable zone, where conditions might allow liquid water to exist. This helps identify planets with conditions suitable for life as we know it.

Exomoons: Exoplanet research also investigates the presence of exomoons (moons orbiting exoplanets), as these moons could play a role in maintaining stable environments for potential life.

Key Discoveries and Implications

Diversity of Planetary Systems: Exoplanet research has revealed a vast array of planetary systems, challenging our previous understanding of planetary formation and dynamics.

Earth-Like Exoplanets: The discovery of "Earth-like" exoplanets in habitable zones raises exciting prospects for finding environments that could potentially support life.

Astrobiology and SETI: Exoplanet research informs the study of astrobiology, the search for life beyond Earth. It guides the search for biosignatures, such as specific atmospheric components, that might indicate the presence of life. It also informs the search for extraterrestrial intelligence (SETI) by identifying potential target systems.

Technological Advancements: Exoplanet research relies heavily on advancements in technology, including high-precision telescopes, sophisticated spectrographs, and data analysis techniques. Missions like the Kepler Space Telescope and the upcoming James Webb Space Telescope are poised to revolutionize our understanding of exoplanets.

Conclusion: Exoplanet research has reshaped our understanding of the cosmos by revealing the prevalence and diversity of planetary systems beyond our solar system. This field expands our knowledge of planetary formation, evolution, and potential habitability, driving humanity's curiosity to explore the universe and seek answers to profound questions about life's existence beyond Earth.

The vastness of the universe doesn't restrict the growth of human knowledge; instead, it propels us toward exploration and innovation. While some regions remain unobservable, our multidisciplinary efforts, technological advancements, theoretical models, and collaborative spirit continue to push the boundaries of our understanding. The perpetual quest for knowledge is an integral part of humanity's journey to unravel the mysteries of the cosmos.

THE SEARCH FOR EXTRATERRESTRIAL LIFE

The possibility of life beyond Earth, either within our observable universe or beyond, has captivated human imagination. Philosophical discussions arise regarding the implications of discovering extraterrestrial life, from the potential for shared knowledge and collaboration to questions about the uniqueness of Earth and humanity.

Philosophical Implications of Discovering Extraterrestrial Life

The prospect of discovering extraterrestrial life has profound philosophical implications that extend beyond the realm of science. It triggers discussions that touch on the nature of life, the uniqueness of Earth, our place in the universe, and the potential for shared knowledge and collaboration. Here are some key philosophical considerations that arise:

UNIVERSALITY OF LIFE

Life's Abundance: The discovery of extraterrestrial life would suggest that life is not a unique phenomenon confined to Earth. This challenges the idea that life on our planet is an exceptional occurrence.

Life's Origins: The existence of life elsewhere could shed light on the origins of life itself, raising questions about whether life arises through similar processes across the universe.

EARTH'S UNIQUENESS AND HUMANITY'S SIGNIFICANCE

The Rare Earth Hypothesis: Some argue that if Earth is the only planet with life, it emphasizes the planet's exceptional qualities. This perspective underscores the significance of humanity's role as caretakers of a unique biosphere.

Human Exceptionalism: The discovery of extraterrestrial

life could challenge anthropocentric views, leading to a reconsideration of humanity's special status in the cosmos.

SHARED KNOWLEDGE AND COLLABORATION

Interstellar Connection: Discovering extraterrestrial life could create an opportunity for interstellar communication and exchange of knowledge. It raises questions about how civilizations might share information and collaborate.

Global Unity: The shared pursuit of contact with extraterrestrial beings might foster a sense of global unity, as the potential encounter could transcend national borders and differences.

ETHICAL AND MORAL CONSIDERATIONS

Respect for Alien Life: The discovery of extra terrestrial life raises questions about our ethical responsibilities toward these potential life forms. How should we approach encounters with life that may be fundamentally different from ours?

Rights and Interactions: If intelligent extraterrestrial species are discovered, discussions may arise regarding their rights and how we should interact with them, especially if they are less technologically advanced.

SOCIETAL AND RELIGIOUS IMPACTS

Cultural Paradigm Shifts: The discovery of extraterrestrial

life could challenge religious beliefs and cultural narratives that place humanity at the center of creation. Societies might need to adapt to new perspectives.

Ethical Implications: Religious and ethical frameworks may need to address the moral implications of interactions with beings that don't share our cultural and religious backgrounds.

Limits of Human Knowledge

Cosmic Significance: Discovering extraterrestrial life highlights our limited understanding of the cosmos. It underscores the vastness of the universe and the potential for even more profound revelations.

In essence, the philosophical discussions surrounding the discovery of extraterrestrial life touch on fundamental aspects of our identity, our place in the universe, and the implications of encountering other intelligent beings. As we contemplate these possibilities, we are prompted to reevaluate our beliefs, values, and how we perceive ourselves in the cosmic context.

Questions of Existence and Reality

The vastness of the universe raises philosophical questions about existence, reality, and the nature of the cosmos itself. Debates about the origin of the universe, the existence of multiple universes (multiverse theory), and the underlying fabric of reality gain new dimensions when placed in the context of the unobservable regions.

Debates About the Origin of the Universe and the Multiverse Theory

Origin of the Universe: The debates surrounding the origin of the universe revolve around understanding how the cosmos came into existence. The prevailing scientific theory, the Big Bang theory, posits that the universe began as an incredibly dense and hot singularity, expanding and evolving into its current state over billions of years. This theory is supported by various observations, such as the cosmic microwave background radiation.

Multiverse Theory: The multiverse theory suggests that our universe is just one of many universes that exist within a larger multiverse. This concept arises from the study of cosmology, quantum mechanics, and string theory. According to some interpretations, the vastness of the cosmos and the underlying fabric of reality could accommodate a multitude of separate universes, each with its own physical laws and conditions.

IMPLICATIONS FOR THE UNDERLYING FABRIC OF REALITY

1. New Dimensions of Reality: The concept of the multiverse adds new dimensions to our understanding of reality. It challenges the notion that our universe is the sole reality, proposing the existence of other realms with different physical constants, forces, and conditions.

2. Unobservable Regions and the Multiverse: The multiverse theory confronts the challenge of unobservable regions. If multiple universes exist, some might be beyond our observational capabilities due to their separation by vast cosmic distances or distinct laws of physics. This introduces an added layer of complexity to comprehending the full extent of reality.

3. Unifying Cosmological Theories: The concept of the multiverse has been proposed as a potential solution to various cosmological puzzles, such as the fine-tuning of physical constants. It suggests that the specific conditions of our universe may be a result of the larger multiverse's diversity.

4. Philosophical and Theoretical Questions: Debates about the multiverse raise philosophical questions about the nature of reality, existence, and the limits of scientific inquiry. Some critics argue that the multiverse is speculative and unobservable, questioning its scientific validity.

5. Challenges to Understanding Reality: The existence of multiple universes adds complexity to our attempts to understand the underlying fabric of reality. It challenges our intuitive grasp of the cosmos and the limitations of human cognition when dealing with concepts that stretch beyond our direct experiences.

Conclusion: Debates about the origin of the universe and the multiverse theory have profound implications for our understanding of the underlying fabric of reality. The exploration of these concepts stretches the boundaries of our current scientific and philosophical knowledge, inviting us to contemplate the nature of existence, the limits of human comprehension, and the interconnectedness of all that may lie beyond the observable regions of the universe.

ETHICAL CONSIDERATIONS

The philosophical implications extend to ethical considerations. As we explore space and potentially encounter other forms of life, questions about our responsibilities as stewards of Earth, our impact on other civilizations, and the ethical treatment of potential extraterrestrial life emerge.

Philosophical Implications and Ethical Considerations of Space Exploration and Extraterrestrial Life

The exploration of space and the potential for encounters with other forms of life give rise to profound philosophical and ethical questions that challenge our understanding of humanity's place in the cosmos. These considerations extend to our responsibilities as stewards of Earth, our potential impact on other civilizations, and the ethical treatment of potential extraterrestrial life.

1. Stewardship of Earth: The realization of Earth's uniqueness and fragility in the vast cosmos prompts discussions about our ethical duty to care for our home planet. The concept of the "Overview Effect" suggests

that seeing Earth from space fosters a sense of unity and responsibility for the planet's well-being.

2. Interactions with Other Civilizations: The possibility of contact with extraterrestrial civilizations forces us to consider how we might ethically engage with beings that possess different cultural, social, and technological backgrounds. This calls for a universal framework that promotes understanding and cooperation.

3. Potential Impact on Other Civilizations: Our actions in space could inadvertently affect other civilizations. Transmitting signals, introducing microbes on spacecraft, or altering the environment of celestial bodies may have consequences for potential life forms elsewhere. Ethical considerations emphasize caution to avoid unintended harm.

4. Ethical Treatment of Extraterrestrial Life: The potential existence of extraterrestrial life raises questions about how we should treat these beings. Are we morally obligated to respect their autonomy and well-being? Ethical frameworks developed on Earth may need to be extended to interactions with other sentient entities.

5. The Precautionary Principle: The precautionary principle suggests that when faced with uncertain risks, we should err on the side of caution to prevent irreversible harm. This principle applies to space exploration, where our activities could have far-reaching consequences for other civilizations or environments.

6. Ethical Universalism: Ethical discussions related to space exploration can lead to reflections on universal ethical principles. If ethical considerations apply to interactions with potential extraterrestrial life, they might also serve as a guide for interactions between humans and other forms of life on Earth.

7. Ethical Dilemmas in Cosmic Colonization: The prospect of establishing human colonies on other planets introduces ethical dilemmas. Questions about property rights, governance, cultural preservation, and sustainability arise as we extend human presence beyond Earth.

8. Cultural and Moral Pluralism: Encountering extraterrestrial civilizations could challenge our own moral and cultural norms. Navigating the diversity of ethical systems and reconciling potential differences in values becomes essential.

Conclusion: The philosophical implications and ethical considerations stemming from space exploration and potential encounters with other forms of life push us to reevaluate our place in the universe and our responsibilities as inhabitants of Earth. Ethical considerations extend from our interactions with other civilizations to our role as stewards of the planet and the universe. The ongoing exploration of space invites us to explore our ethical values, reflect on our impact on other life forms, and forge a framework that respects the diversity of life and civilizations beyond our home planet.

In essence, contemplating the universe's vastness propels us into realms of philosophical introspection and exploration. It encourages us to grapple with questions about existence, meaning, and our place in the grand tapestry of cosmic existence. These philosophical considerations not only deepen our understanding of the universe but also prompt us to reflect on human experience, our responsibilities, and the collective journey of exploration and discovery.

Chapter 5

The Miracles of Biology and Medicine

Exploring the Intricacies of Human Biology and the Wonders of Life

The human body is a masterpiece of biological engineering, a complex symphony of cells, tissues, and organs working in harmonious synchrony. In this chapter, we take you into the astonishing intricacies of human biology and the remarkable wonders that define life itself.

1. THE BLUEPRINT OF LIFE: DNA AND GENETICS

We explore the fundamental unit of life, DNA, and how its unique code holds the blueprint for every living organism. Unraveling the intricacies of genetics, we examine how DNA replication, mutations, and gene expression shape our traits and contribute to the breathtaking diversity of life on Earth.

2. THE DANCE OF CELLS: FROM MICROSCOPIC TO MACROSCOPIC

From the microscopic realm of cells to the macroscopic complexity of organs, the human body is a living testament to the elegance of biological systems. We journey through the intricacies of cellular processes, the formation of tissues and organs, and the interplay of biological systems that sustain our existence.

3. THE SYMPHONY OF LIFE: HOMEOSTASIS AND ECOSYSTEMS

Life is a delicate balance of dynamic processes that maintain equilibrium within the body and its environment. We delve into the concept of homeostasis—the body's ability to regulate its internal conditions—and explore how ecosystems, from rainforests to coral reefs, exemplify the interconnectedness of all life forms on Earth.

Medical Breakthroughs and the Conquest of Diseases

Throughout history, the pursuit of medical knowledge has led to transformative breakthroughs that have alleviated human suffering and prolonged life. In this section, we celebrate the remarkable achievements of medical science and the conquest of diseases that have shaped our species' progress.

1. Vaccines: Triumph over Pathogens

The development of vaccines stands as a monumental achievement in medical science. We delve into the history of vaccination, from Edward Jenner's smallpox inoculation to the modern triumphs against diseases such as polio, measles, and COVID-19. Vaccines have not only saved countless lives but have also transformed the landscape of public health.

2. Surgical Innovations: From Scalpels to Robotics

Advancements in surgical techniques and technologies have revolutionized the way we approach medical interventions. We explore the evolution of surgery, from ancient practices to minimally invasive procedures and robotic-assisted surgeries. These innovations have increased precision, reduced recovery times, and expanded the boundaries of what medical science can achieve.

3. Genetics and Personalized Medicine

The mapping of the human genome has opened the door to a new era of personalized medicine. We examine how genetic insights are driving tailored treatment plans for individuals, addressing genetic predispositions to diseases, and optimizing therapeutic outcomes. The promise of precision medicine heralds a future where treatments are finely tuned to an individual's unique genetic makeup.

Ethical Dilemmas and the Future of Genetic Engineering and Regenerative Medicine

As scientific knowledge advances, ethical considerations become ever more vital. This section delves into the ethical complexities surrounding genetic engineering, regenerative medicine, and the profound impact of these technologies on society and the human experience.

1. GENETIC ENGINEERING: POSSIBILITIES AND PERILS

The ability to manipulate genes raises both remarkable possibilities and ethical dilemmas. We explore the potential of genetic engineering to cure genetic diseases, enhance human traits, and even alter the course of evolution. Amidst the promise, we also discuss concerns about unintended consequences, genetic inequality, and the ethical boundaries we must navigate.

2. REGENERATIVE MEDICINE: FROM HEALING TO ENHANCEMENT

Regenerative medicine offers the potential to restore damaged tissues and organs, revolutionizing healthcare. We explore breakthroughs such as stem cell therapies and tissue engineering and discuss the ethical questions arising from the intersection of healing and enhancement. As we push the boundaries of human capabilities, we must confront the moral implications of transforming the human body.

3. BIOETHICS AND THE FUTURE

The intertwining of science, ethics, and human values shapes the trajectory of our species. We delve into the realm of bioethics, examining the principles that guide our decisions about emerging medical technologies. As we stand at the crossroads of possibility, we ponder how society can strike a balance between scientific advancement, human values, and responsible innovation.

In the realm of biology and medicine, we witness the remarkable dance of life, from the intricate workings of cells to the towering achievements of medical science. As we navigate the complexities of medical breakthroughs and technological wonders, we are called upon to consider the profound ethical questions that accompany our journey. Join us as we unravel the miracles of human biology, explore the frontiers of medical innovation, and contemplate the boundless potential—and responsibilities—that lie on the horizon of genetic engineering and regenerative medicine.

Exploring the Intricacies of Human Biology and the Wonders of Life

The human body is a masterpiece of biological engineering, a complex symphony of cells, tissues, and organs working in harmonious synchrony. In this chapter, we investigate the astonishing intricacies of human biology and the remarkable wonders that define life itself. In the realm of biology and medicine, we witness the remarkable dance of life, from the intricate workings of cells to the towering achievements of medical science. As we navigate the complexities of medical breakthroughs and technological wonders, we are also called upon to consider the profound ethical questions that accompany our journey. Stay with us as we unravel the miracles of human biology, explore the frontiers of medical innovation, and contemplate the boundless potential—and responsibilities—that lie on the horizon of genetic engineering and regenerative medicine.

THE BLUEPRINT OF LIFE: DNA AND GENETICS

We explore the fundamental unit of life, DNA, and how its unique code holds the blueprint for every living organism. Unraveling the intricacies of genetics, we examine how DNA replication, mutations, and gene expression shape our traits and contribute to the breathtaking diversity of life on Earth.

DNA (deoxyribonucleic acid) is the fundamental unit of life and serves as the genetic blueprint for all living organisms. It contains the instructions necessary for an organism's growth, development, functioning, and reproduction. The unique code within DNA governs the characteristics that make each living being distinct.

STRUCTURE OF DNA

DNA is composed of two long chains made up of nucleotides, which are the building blocks of DNA. Each nucleotide consists of a phosphate group, a sugar molecule (deoxyribose), and one of four nitrogenous bases: adenine

(A), thymine (T), cytosine (C), and guanine (G). The arrangement of these nitrogenous bases along the DNA strands forms the genetic code.

DNA Replication

Before a cell divides, its DNA needs to be replicated to ensure that each daughter cell receives a complete set of genetic information. During replication, the two DNA strands unwind, and each serves as a template for the creation of a new complementary strand. This process is highly accurate due to the specificity of base pairing (A with T, and C with G). DNA polymerase enzymes facilitate the synthesis of new strands, resulting in two identical DNA molecules.

Mutations

While DNA replication is accurate, occasional errors—mutations—can occur. Mutations are changes in the DNA sequence, and they can be caused by various factors, including radiation, chemicals, or replication errors. Some mutations have no significant impact, while others can lead to changes in an organism's traits. Mutations are essential for evolutionary processes, introducing genetic diversity that can provide advantages or disadvantages in different environments.

Gene Expression

The information encoded in DNA must be translated into

proteins and other molecules that carry out cellular functions. This process is called gene expression. It involves two main stages: transcription and translation.

Transcription: In the cell's nucleus, a portion of the DNA sequence, known as a gene, is transcribed into a molecule called messenger RNA (mRNA). RNA polymerase catalyzes the synthesis of mRNA using DNA as a template.

Translation: The mRNA molecule travels to the cell's cytoplasm, where ribosomes read its code in sets of three nucleotides called codons. Transfer RNA (tRNA) molecules, carrying specific amino acids, and match their anticodons with the mRNA codons. This leads to the assembly of a protein chain, which folds into its functional structure.

GENETIC DIVERSITY AND TRAITS

Variations in the DNA sequence contribute to the breathtaking diversity of life on Earth. Genes encode specific traits that determine an organism's appearance, behavior, and physiological characteristics. Differences in genes and their expression give rise to the vast array of species and the unique individuals within each species.

In, DNA is the exquisite molecule that holds the blueprint of life. Its structure, replication, mutations, and the complex process of gene expression shape our individual traits and collectively contribute to the astonishing diversity of life forms. Through DNA's elegant code, life's complexities unfold, offering a window into the evolutionary processes that have shaped and continue to shape the living world around us.

THE DANCE OF CELLS: FROM MICROSCOPIC TO MACROSCOPIC

From the microscopic realm of cells to the macroscopic complexity of organs, the human body is a living testament to the elegance of biological systems.

FROM CELLS TO SYSTEMS: THE INTRICACIES OF SUSTAINING LIFE

The journey from a single cell to a complex, multicellular organism is a remarkable testament to the intricacies of cellular processes, tissue formation, organ development, and the interplay of biological systems. This journey is orchestrated by an array of precise mechanisms that sustain our existence and enable the functioning of every organ and system in our body.

Cellular Processes: At the heart of life are the fundamental cellular processes that ensure the growth, maintenance, and survival of individual cells. These processes include:

Cell Division: Cells replicate through processes like mitosis (for somatic cells) and meiosis (for sex cells), ensuring the continuity of life and the generation of new cells.

Metabolism: Cells perform chemical reactions to obtain energy from nutrients and build molecules essential for growth and function.

Signal Transduction: Cells communicate with each other through signaling pathways, allowing them to respond to external cues and coordinate actions.

Tissue Formation: Cells with similar functions and structures aggregate to form tissues, which are specialized groups that carry out specific roles in the body. The four primary types of tissues are epithelial, connective, muscle, and nervous tissues. These tissues combine to create organs with distinct functions.

Organ Development: Organs are complex structures formed by multiple tissues working together. Organ development involves intricate processes such as cell differentiation, migration, and pattern formation. Genetic signals guide cells to become specific types and their spatial arrangement results in the formation of functional organs.

Biological Systems: Our body consists of several interdependent biological systems, each responsible for particular functions. These systems include:

Circulatory System: The heart pumps blood, which carries nutrients, oxygen, and waste products throughout the body via blood vessels.

Respiratory System: Oxygen is taken in through the respiratory system and exchanged for carbon dioxide, supporting cellular respiration.

Digestive System: Nutrients are broken down and absorbed

through the digestive system, providing energy and building blocks for cells.

Nervous System: Neurons transmit electrical signals, enabling communication between different parts of the body and regulating responses to stimuli.

Endocrine System: Glands secrete hormones that regulate various physiological processes and maintain homeostasis.

Musculoskeletal System: Muscles provide movement, while bones offer support and protection.

Immune System: Specialized cells defend against pathogens and maintain a healthy internal environment.

Interplay of Systems

The proper functioning of these systems relies on intricate interactions and feedback loops. For example, the circulatory system supplies oxygen and nutrients to tissues, while the respiratory system ensures oxygen intake and carbon dioxide removal. The nervous system coordinates responses to stimuli, while the endocrine system releases hormones that regulate metabolism and growth.

Homeostasis

Throughout these processes, the body maintains homeostasis—the internal balance required for optimal function. This balance involves regulating factors like temperature, pH, and nutrient levels to ensure cells can function properly.

The journey from cells to biological systems showcases the intricacies of sustaining life. Cellular processes, tissue formation, organ development, and the interplay of systems contribute to the functioning of our body. Each component plays a vital role, reflecting the complexity and beauty of the mechanisms that enable us to exist, adapt, and thrive in our dynamic and ever-changing environment.

THE SYMPHONY OF LIFE: HOMEOSTASIS AND ECOSYSTEMS

Life is a delicate balance of dynamic processes that maintain equilibrium within the body and its environment. Here we explore the concept of homeostasis—the body's ability to regulate its internal conditions—and explore how ecosystems, from rainforests to coral reefs, exemplify the interconnectedness of all life forms on Earth.

HOMEOSTASIS: BALANCING THE BODY'S INTERNAL ENVIRONMENT

Homeostasis is a fundamental concept that refers to the body's ability to maintain stable internal conditions despite external changes. It involves a dynamic process of regulation and feedback mechanisms that ensure the body's internal environment remains within a narrow range conducive to optimal functioning. Here's how homeostasis works:

Sensors: Specialized cells and receptors throughout the body detect changes in various parameters like temperature, pH, and nutrient levels.

Control Center: The brain and other regulatory centers process the information received from sensors and compare it to a set point—the desired value for each parameter.

Effector Responses: Based on the comparison, the control center sends signals to effectors—organs or systems that can modify the parameter. These effectors work to bring the parameter back to the set point.

Feedback Mechanisms: Feedback loops are used to regulate the parameter. Negative feedback loops counteract deviations from the set point by initiating responses that bring the parameter back to its optimal level. Positive feedback loops amplify deviations, often with specific purposes like childbirth or blood clotting.

Ecosystem Interconnectedness: Rainforests and Coral Reefs

Ecosystems exemplify the interconnectedness of all life forms on Earth. Two prime examples are rainforests and coral reefs:

Rainforests:

Rainforests are complex ecosystems found in tropical regions. They showcase intricate interactions among various species, plants, animals, and microorganisms. Interconnectedness in rainforests includes:

(I.) Biodiversity: Rainforests house a diverse array of species that rely on each other for food, shelter, and reproduction. The loss of one species can have cascading effects on the entire ecosystem.

(II.) Nutrient Cycling: Plants, animals, and microorganisms in rainforests contribute to nutrient cycling. Decomposers break down organic matter, releasing nutrients that are taken up by plants, sustaining their growth.

(III.)Symbiotic Relationships: Rainforests are home to numerous symbiotic relationships. For instance, certain plants rely on specific insects for pollination, while animals may disperse seeds, benefiting both parties.

Coral Reefs:

(I.)Coral reefs are marine ecosystems built by coral colonies. They demonstrate intricate connections between organisms and their environment:

(II.)Coral Symbiosis: Coral polyps have a mutualistic relationship with photosynthetic algae called zooxanthellae. The algae provide energy through photosynthesis, while the polyps offer shelter and nutrients.

(III.)Biodiversity Hotspots: Coral reefs host a rich variety of marine life. Fish and other species depend on the complex structure of the reef for shelter, hunting, and reproduction.

Food Web: Reefs feature intricate food webs. Predatory fish control populations of herbivores, which in turn prevent algae overgrowth that can harm coral.

In both rainforests and coral reefs, the delicate balance of interactions and interdependencies sustains the overall health and function of the ecosystem. When disruptions occur, whether through human activities or natural events, the consequences can reverberate throughout the entire ecosystem.

In conclusion, the concept of homeostasis applies not only to individual organisms but also to the broader context of ecosystems. Just as our bodies strive to maintain equilibrium, ecosystems rely on interconnected relationships to maintain balance and sustain life. The intricate web of interactions and dependencies within ecosystems, whether in rainforests or coral reefs, underscores the profound interconnectedness of all life forms on Earth and high- lights the delicate nature of our planet's biodiversity and health.

Medical Breakthroughs and the Conquest of Diseases

Throughout history, the pursuit of medical knowledge has led to transformative breakthroughs that have alleviated human suffering and prolonged life. In this section, we celebrate the remarkable achievements of medical science and the conquest of diseases that have shaped our species' progress.

The Evolution of Vaccination: A Triumph of Public Health (Vaccines: Triumph over Pathogens)

Vaccination has been one of the most significant and impactful advances in the field of medicine, playing a pivotal role in preventing and controlling the spread of infectious diseases. From the pioneering efforts of Edward Jenner to the modern successes against diseases like polio, measles, and COVID-19, vaccines have saved countless lives and transformed the landscape of public health.

Edward Jenner and Smallpox Vaccination: In 1796, English physician Edward Jenner introduced the concept of vaccination by using cowpox to protect against smallpox. He observed that milkmaids who contracted cowpox—a mild disease—were subsequently immune to smallpox. He conducted an experiment by inoculating a boy with cowpox and then exposing him to small pox, finding that the boy did not develop the disease. This marked the beginning of immunization and laid the foundation for later vaccine development.

Vaccine Development and Public Health Triumphs

Throughout history, vaccine development has tackled a range of deadly diseases:

Polio: Jonas Salk's inactivated polio vaccine and Albert Sabin's oral attenuated vaccine led to the near eradication of polio. The Global Polio Eradication Initiative has drastically reduced cases worldwide.

Measles: The introduction of the measles vaccine in the 1960s significantly reduced the incidence of this highly contagious disease.

COVID-19: In an extraordinary example of modern vaccine development, multiple COVID-19 vaccines were developed and authorized for emergency use within a year of the pandemic's onset. These vaccines have played a crucial role in controlling the spread of the virus and reducing severe illness and death.

IMPACT ON PUBLIC HEALTH

Vaccines have brought about transformative changes in public health as follows:

- Disease Prevention: Vaccines prevent illness, disability, and death from a variety of infectious diseases. They create immunity without causing the disease itself.

- Herd Immunity: Vaccination contributes to herd immunity, where a sufficiently high percentage of a population becomes immune, indirectly protecting those who cannot be vaccinated due to medical conditions.

- Eradication and Elimination: Vaccines have led to the

eradication of smallpox and the near eradication of polio. They are also being used to eliminate diseases like measles and rubella in certain regions.

- Global Health Equity: Vaccination campaigns, especially in low-income countries, have been instrumental in improving health equity and reducing health disparities.

- Preparedness for Emerging Diseases: The rapid development of vaccines against new threats, such as COVID-19, demonstrates the scientific and technological advances made in the field.

- Economic Impact: Vaccination reduces healthcare costs, prevents lost productivity due to illness, and saves lives, contributing to a healthier and more productive society.

ONGOING CHALLENGES AND FUTURE PROSPECTS

Despite the successes, challenges remain, including vaccine hesitancy, equitable distribution, and addressing emerging diseases. Researchers continue to work on innovative vaccine technologies, such as mRNA vaccines used for COVID-19, which may have applications beyond infectious diseases.

In conclusion, the history of vaccination is a testament to human ingenuity, scientific progress, and the power of collective efforts in public health. Vaccines have saved countless lives, prevented suffering, and transformed societies. From Jenner's smallpox inoculation to the current fight against COVID-19, vaccines stand as a beacon of hope and a testament to the potential of medicine to shape a healthier future for humanity.

The Evolution of Antibiotics: A Medical Milestone

The discovery and development of antibiotics have revolutionized modern medicine, saving countless lives and transforming the landscape of public health. Here's an exploration of the history of antibiotics, their evolution, impact, challenges, and ongoing significance:

1. INCEPTION OF ANTIBIOTICS:

Penicillin: The era of antibiotics began in 1928 when Scottish bacteriologist Alexander Fleming discovered penicillin. He noticed that a mold called Penicillium notatum produced a substance that killed a wide range of bacteria. This marked the first recorded use of antibiotics to combat bacterial infections.

2. ANTIBIOTIC DEVELOPMENT AND IMPACT:

World War II: The potential of penicillin for treating infections became evident during World War II, when it played a crucial role in saving the lives of wounded soldiers.

Mass Production: In the 1940s, a collaborative effort led to the mass production of penicillin, making it widely available.

Antibiotic Era: The discovery of other antibiotics followed, including streptomycin, tetracycline, and erythromycin, which expanded the arsenal against various bacterial infections.

3. Transforming Public Health:

Major Medical Advancement: Antibiotics transformed medicine by providing effective treatments for bacterial infections, reducing mortality rates from previously deadly diseases.

Surgical Procedures: The availability of antibiotics enabled safer surgical procedures, as infections that were once fatal complications could be treated.

4. Challenges and Concerns:

Antibiotic Resistance: The overuse and misuse of antibiotics led to the development of antibiotic-resistant bacteria. Resistant strains like MRSA (Methicillin-resistant Staphylococcus aureus) pose a significant challenge to modern medicine.

Discovery Gap: After the initial surge of antibiotic discovery, there was a decline in the development of new antibiotics due to challenges in research and development.

5. Modern Advances and Future Prospects:

Combating Resistance: Researchers are exploring strategies to combat antibiotic resistance, including developing new antibiotics, improving stewardship programs, and finding alternative treatments.

Phage Therapy: Bacteriophage therapy, which uses viruses to target specific bacteria, is being investigated as an alternative to antibiotics.

Precision Medicine: Advances in genomics and personalized medicine are leading to the development of antibiotics tailored to specific bacterial strains.

6. ONE HEALTH APPROACH:

Human and Animal Health: Antibiotic use in livestock and agriculture contributes to the emergence of antibiotic-resistant bacteria. A One Health approach considers the inter- connectedness of human, animal, and environmental health.

7. GLOBAL CHALLENGES AND COOPERATION:

Global Health Threat: Antibiotic resistance is a global health threat that requires international cooperation, surveillance, and coordinated efforts to prevent the spread of resistant bacteria.

Antibiotics have had a profound impact on public health by revolutionizing the treatment of bacterial infections. However, the rise of antibiotic-resistant bacteria underscores the importance of responsible antibiotic use, innovative research, and a collaborative approach to address the challenges posed by antibiotic resistance. The history of antibiotics serves as a reminder of the power of scientific discovery to improve human well-being and the ongoing need for vigilance and innovation to address emerging health threats.

Surgical Innovations: From Scalpels to Robotics

Advancements in surgical techniques and technologies have revolutionized the way we approach medical interventions. We explore the evolution of surgery, from ancient practices to minimally invasive procedures and robotic-assisted surgeries. These innovations have increased precision, reduced recovery times, and expanded the boundaries of what medical science can achieve.

THE EVOLUTION OF SURGERY: ADVANCEMENTS IN MEDICAL SCIENCE

The history of surgery is a journey marked by innovation, experimentation, and remarkable breakthroughs that have transformed medical practices and patient outcomes. From ancient civilizations to the modern era, surgery has evolved from crude procedures to sophisticated techniques, including minimally invasive and robotic-assisted surgeries.

1. Ancient Practices:

Early Techniques: Ancient civilizations like the Egyptians, Greeks, and Romans practiced surgery using basic tools and techniques. Procedures were often crude and associated with high risks of infection and mortality.

Innovations: The ancient Greeks introduced concepts of surgical hygiene and anatomy, while Indian and Arab cultures developed methods of wound closure and surgical instruments.

2. Renaissance and Enlightenment:

Anatomy and Dissection: The Renaissance period witnessed a revival of interest in anatomy and dissection, leading to a better understanding of human anatomy and the foundation for modern surgery.

Anesthesia: The discovery of anesthesia in the 19th century revolutionized surgery by eliminating pain and allowing for more complex procedures.

3. Modern Surgical Techniques:

Aseptic Techniques: The introduction of aseptic techniques and sterilization in the 19th century greatly reduced the risk of infection during surgery.

Advances in Imaging: X-rays, computed tomography (CT), and magnetic resonance imaging (MRI) revolutionized diagnosis and enabled better pre-operative planning.

4. Minimally Invasive Surgery:

Laparoscopy: In the 20th century, laparoscopic surgery emerged, using small incisions and specialized instruments to perform procedures inside the body. This led to reduced pain, shorter hospital stays, and faster recovery times.

Endoscopy: Similar principles were applied to endoscopy, allowing visualization and treatment of internal structures through natural openings or small incisions.

5. Robotic-Assisted Surgery:

Robotic Systems: In recent years, robotic-assisted surgery has gained prominence. Robots with precise movements and magnified visuals enable surgeons to perform complex procedures with enhanced dexterity.

Da Vinci System: The Da Vinci Surgical System, for example, allows for minimally invasive surgery with greater precision and less tissue damage.

6. Benefits and Implications:

Precision and Accuracy: Minimally invasive and robotic-assisted surgeries offer increased precision, allowing surgeons to navigate intricate anatomy and perform delicate procedures.

Reduced Recovery Time: Smaller incisions and reduced trauma to surrounding tissues lead to shorter recovery times, decreased pain, and lower risk of complications.

Expanded Possibilities: Advanced techniques enable surgeries that were once considered high-risk or impossible, pushing the boundaries of medical science.

7. Challenges and Future Directions:

Training: Surgeons require specialized training to use robotic systems effectively, and challenges such as high costs and limited access to technology persist.

Ethical Considerations: As technology evolves, ethical considerations arise, such as patient consent and the role of technology in decision-making.

In conclusion, the evolution of surgery is a testament to human innovation, scientific discovery, and the relentless pursuit of improved patient outcomes. From ancient practices to modern robotic-assisted surgeries, each advancement has built upon the lessons of the past. Minimally invasive and robotic techniques have revolutionized surgical care, offering increased precision, shorter recovery times, and expanded possibilities for patients. As technology continues to evolve, the field of surgery will undoubtedly see further transformative changes that enhance patient care and redefine the boundaries of medical achievement.

Genetics and Personalized Medicine

The mapping of the human genome has opened the door to a new era of personalized medicine. We examine how genetic insights are driving tailored treatment plans for individuals, addressing genetic predispositions to diseases, and optimizing therapeutic outcomes. The promise of precision medicine heralds a future where treatments are finely tuned to an individual's unique genetic map.

PERSONALIZED MEDICINE: UNLOCKING THE POWER OF THE HUMAN GENOME

The mapping of the human genome has ushered in a new era of personalized medicine, where our understanding of genetics is transforming how medical care is delivered. This revolution is fueled by the insights gained from decoding the human genome and the ability to tailor treatments to an individual's unique genetic makeup.

1. The Human Genome Project:

Genetic Blueprint: The Human Genome Project, completed in 2003, mapped the entire human genome, identifying all genes and their functions. This monumental effort provided a comprehensive understanding of our genetic makeup.

2. Genetic Insights and Personalized Treatment:

Targeted Therapies: Genetic insights allow doctors to identify specific genetic mutations associated with diseases.

Targeted therapies are designed to address these mutations, improving treatment efficacy, and minimizing side effects.

Pharmacogenomics: Understanding how genes affect an individual's response to medications has led to pharmacogenomic approaches, optimizing drug selection and dosages.

3. Genetic Predisposition and Disease Prevention:

Risk Assessment: Genetic testing helps identify genetic predispositions to diseases. Early detection of genetic markers allows for proactive interventions, lifestyle modifications, and screening.

Cancer Risk: Genetic testing can identify individuals with a higher risk of developing certain types of cancer, leading to early detection and preventive measures.

4. Personalized Treatment Plans:

Tailored Therapies: Personalized medicine considers an individual's genetic makeup, health history, and lifestyle to create treatment plans tailored to their unique needs.

Cancer Treatment: Oncologists analyze the genetic profile of tumors to select treatments with the highest likelihood of success and minimal side effects.

5. Precision Medicine and the Future:

Enhanced Outcomes: By optimizing treatments based on genetic information, precision medicine aims to improve therapeutic outcomes, reduce adverse effects, and increase patient satisfaction.

Rare Diseases: Genetic insights are vital in diagnosing and treating rare diseases that often have elusive origins and limited treatment options.

Population Health: Aggregate genetic data can reveal patterns of disease prevalence and guide public health strategies.

6. Challenges and Considerations:

Data Privacy: The use of genetic data raises concerns about privacy, security, and ethical use.

Access and Equity: Ensuring equitable access to personalized medicine and addressing disparities in genetic testing and treatment are critical.

7. Collaboration and Research:

Genomic Databases: Sharing genetic data globally facilitates research and accelerates discoveries.

Drug Development: Genetic insights guide drug development, leading to more effective therapies.

In summary, the mapping of the human genome has paved

the way for personalized medicine, transforming healthcare by tailoring treatments to individual genetic profiles. This approach optimizes treatment outcomes, minimizes adverse effects, and addresses genetic predispositions to diseases. As research advances and technology evolves, precision medicine promises a future where medical care is finely tuned to an individual's unique genetic makeup, revolutionizing the way we prevent, diagnose, and treat diseases. Top of Form

Ethical Dilemmas and the Future of Genetic Engineering and Regenerative Medicine

As scientific knowledge advances, ethical considerations become ever more vital. This section delves into the ethical complexities surrounding genetic engineering, regenerative medicine, and the profound impact of these technologies on society and the human experience.

1. Genetic Engineering: Possibilities and Perils

The ability to manipulate genes raises both remarkable possibilities and ethical dilemmas. We explore the potential of genetic engineering to cure genetic diseases, enhance human traits, and even alter the course of evolution. Amidst the promise, we also delve into concerns about unintended consequences, genetic inequality, and the ethical boundaries we must navigate.

GENETIC ENGINEERING: PROMISE AND ETHICAL DILEMMAS

Genetic engineering holds immense potential to reshape our understanding of biology and medicine, with the ability to cure genetic diseases, enhance human traits, and even influence the course of evolution. While these advancements offer transformative possibilities, they also raise complex ethical considerations and concerns.

1. Curing Genetic Diseases:

Gene Therapy: Genetic engineering can introduce functional genes into cells to correct genetic mutations causing diseases like cystic fibrosis or sickle cell anemia.

CRISPR-Cas9: This revolutionary tool enables precise editing of DNA, potentially correcting disease-causing mutations.

2. Enhancing Human Traits:

Designer Babies: Genetic engineering might allow us to select or enhance specific traits in unborn children, such as intelligence or physical attributes.

Ethical Concerns: Enhancements could lead to unequal access, societal pressure to conform to certain traits, and the commodification of human traits.

3. Altering Evolution:

Gene Drive: Using gene editing to engineer organisms with traits that spread through populations, potentially altering the genetic makeup of entire species.

Environmental Impact: Unintended ecological consequences could arise if engineered organisms affect ecosystems.

4. Unintended Consequences:

Off-Target Effects: Genetic editing techniques like CRISPR can have unintended mutations or off-target effects that may lead to unforeseen outcomes.

Long-Term Effects: The consequences of genetic modifications may not be immediately evident and could have long- term implications.

5. Genetic Inequality:

Access Disparities: Advanced genetic therapies may not be universally accessible, leading to genetic inequality and exacerbating existing disparities in healthcare.

Economic Divide: Only those with resources might have access to genetic enhancements, potentially creating an economic divide based on genetic makeup.

6. Ethical Boundaries:

Respect for Autonomy: Individuals should have the autonomy to make decisions about their genetic makeup.

Playing Nature's Role: Altering human traits or the course of evolution raises concerns about our role in tampering with the natural order.

Consent and Informed Choice: Ethical genetic engineering requires informed consent and a comprehensive understanding of potential risks and benefits.

7. Regulatory Frameworks:

Global Cooperation: Developing international regulations for genetic engineering is crucial to ensure responsible and ethical practices.

Oversight and Accountability: Ethical considerations demand rigorous oversight of research, trials, and applications.

The potential of genetic engineering to cure diseases, enhance traits, and influence evolution is therefore both fascinating and ethically challenging. While advancements hold the promise of improved health and well-being, they also necessitate careful consideration of ethical boundaries, unintended consequences, and the potential for genetic inequality. Striking a balance between scientific progress and ethical responsibility is crucial as we navigate the complexities of genetic engineering and its impact on humanity and the natural world.

Stem cell therapies and tissue engineering hold immense promise for healing and potentially enhancing human capabilities. The ethical questions they raise underscore the need for thoughtful deliberation, responsible research, and clear regulatory frameworks. As we navigate the intersection of healing and enhancement, it is imperative to address the moral implications, ensure equitable access, and strike a balance between scientific progress and ethical considerations, ultimately shaping a future that respects human values and advances medical science. Through regulatory frameworks, collaborative efforts, and open dialogues, scientists strive to create a future where stem cell therapies uphold human values, respect individual choices, and not just advance medical science but do so responsibly.

Addressing Ethical Questions in Stem Cell Therapy: Balancing Healing and Enhancement

The field of science is actively engaging with the complex ethical questions arising from the intersection of healing and enhancement through stem cell therapy. This engagement involves considering moral implications, ensuring equitable access, and striking a balance between scientific progress and ethical considerations. Let's explore how these aspects are being addressed:

1. Moral Implications:

Therapeutic Focus: The scientific community emphasizes using stem cell therapies primarily for therapeutic purposes, such as treating diseases and injuries, rather than pursuing enhancements.

Ethics Committees: Research institutions often establish ethics committees to evaluate proposed studies and ensure they align with ethical guidelines.

2. Equitable Access:

Affordability: Ensuring that stem cell therapies are accessible to a diverse range of patients is essential. Efforts are made to minimize costs and explore funding options.

Global Health Initiatives: Collaborative projects seek to make stem cell therapies available in low-and middle-income countries, reducing disparities in access.

3. Striking a Balance:

Informed Consent: Clear and comprehensive informed consent procedures ensure that patients understand the potential benefits, risks, and uncertainties of stem cell therapies.

Regulatory Oversight: Governments and international bodies develop regulations that prevent the misuse of stem cell therapies for enhancement purposes. Examples:

Stem Cell Therapy for Parkinson's Disease:

Healing Focus: Researchers are using stem cells to replace damaged dopamine-producing neurons in patients with Parkinson's disease, aiming to alleviate symptoms and improve quality of life.

Tissue Regeneration for Spinal Cord Injuries:

Therapeutic Intent: Stem cell therapies are being investigated to repair damaged spinal cords and restore mobility and function to individuals with spinal cord injuries.

GLOBAL COLLABORATIONS

World Health Organization (WHO): WHO provides guidelines and ethical considerations for stem cell research and therapies to ensure global standards are upheld.

International Stem Cell Registries: Collaborative efforts create databases to track stem cell therapies' outcomes, enhancing transparency and accountability.

ETHICAL DIALOGUES AND PUBLIC ENGAGEMENT

Bioethics Discussions: Scientific conferences and academic journals regularly address ethical issues in stem cell therapies, fostering critical discussions.

Public Education: Engaging the public through educational initiatives ensures informed decision-making about medical interventions.

Balancing Human Values and Progress: The key to shaping a responsible future lies in maintaining a focus on human values:

Ethical Reflection: Constant evaluation of the potential societal impacts of stem cell therapies helps prevent unintended consequences.

Empowerment: Ensuring patients understand the implications of choosing stem cell therapies enables them to make informed decisions aligned with their values.

Bioethics: Balancing Scientific Advancement and Human Values

In the realm of bioethics, guiding principles help us navigate the complex landscape of emerging medical technologies, ensuring that scientific advancement aligns with human values and responsible innovation. These principles provide a framework for ethical decision-making and help strike a balance between progress and societal well-being.

1. Autonomy:

Respect for Individuals: The principle of autonomy emphasizes the right of individuals to make informed decisions about their own bodies and medical treatments.

Informed Consent: Emerging medical technologies require transparent communication about potential risks, benefits, and alternatives to empower individuals to make autonomous choices.

2. Beneficence:

Promoting Well-being: The principle of beneficence focuses on maximizing benefits for individuals and society while minimizing potential harm.

Risk-Benefit Assessment: Before implementing emerging technologies, a thorough evaluation of potential benefits and risks ensures that innovations lead to overall positive outcomes.

3. Nonmaleficence:

Do No Harm: Nonmaleficence underscores the importance of avoiding harm to individuals and society through medical interventions.

Precautionary Approach: Society should take a cautious approach to emerging technologies, ensuring that unforeseen negative consequences are minimized.

4. Justice:

Fair Distribution: The principle of justice highlights the need for equitable access to emerging medical technologies, regardless of socioeconomic status or other factors.

Addressing Inequities: Ensuring access to all segments of society promotes fairness and prevents further disparities.

5. Respect for Dignity:

Human Worth: Every individual's intrinsic worth and rights should be upheld, even in the face of rapid technological advancements.

Preventing Exploitation: Technologies should be developed and implemented in ways that respect the dignity of individuals and avoid exploitative practices.

6. Prudent Stewardship:

Responsible Innovation: Prudent stewardship emphasizes the need for thoughtful oversight of emerging technologies to prevent unintended consequences.

Regulatory Frameworks: Robust regulatory systems and ethical guidelines help ensure that innovations are developed and deployed responsibly.

STRIKING A BALANCE:

1. Ethical Reflection and Dialogue:

Encouraging open discussions among scientists, ethicists, policymakers, and the public allows for comprehensive examination of the ethical implications of emerging technologies.

Deliberation ensures that decisions are informed by diverse perspectives and ethical considerations.

2. Public Engagement:

Engaging the public in decision-making processes fosters awareness, understanding, and shared responsibility for the ethical impact of new technologies.

Public input ensures that societal values influence the direction of scientific advancements.

3. Collaborative Efforts:

International collaborations create a global perspective on ethical issues, promoting consistent standards and responsible innovation.

Shared best practices and guidelines prevent unethical practices from proliferating in different regions.

4. Continuous Evaluation:

Regular assessment of the ethical implications of emerging technologies ensures that ethical considerations evolve alongside scientific progress.

Flexibility allows adjustments in response to new insights and changing societal needs.

In conclusion, bioethics guides our decisions about emerging medical technologies by providing a set of principles that prioritize human values, responsible innovation, and societal well-being. By upholding principles of autonomy, beneficence, nonmaleficence, justice, dignity, and prudent stewardship, society can strike a delicate balance between scientific advancement and ethical considerations, ensuring that progress enriches lives without compromising human values.

Part Three

Humanity's Potential and Future

Chapter 6

The Power of Technology

The Impact of Technological Advancements on Society

Technology has woven itself into the fabric of human existence, shaping our societies, economies, and daily lives. In this chapter, we look into the profound impact of technological advancements on our world and examine how they have reshaped the course of history. Balancing Progress and Responsibility in a Technology-Driven World is crucial. The power of technology comes with a responsibility to use it wisely. In this section, we grapple with the ethical and societal implications of our rapidly evolving technological landscape.

Image of technological advancements

Technological Evolution: From Fire to the Internet

We trace the trajectory of technological progress, from the discovery of fire to the invention of the wheel, the printing press, and beyond. Each milestone in human innovation has driven societal transformations, enabling the exchange of knowledge, the growth of economies, and the evolution of cultural norms. Human history is a tapestry woven with the threads of technological innovation. From the mastery of fire, which provided warmth, protection, and the ability to cook, to the invention of the wheel, which revolutionized transportation and trade, each breakthrough has propelled human society forward. The printing press democratized knowledge dissemination, enabling the spread of ideas across continents. The industrial revolution transformed economies and manufacturing, while the digital revolution unleashed the power of information exchange on a global scale.

The Digital Revolution: Transforming Communication and Information

The advent of computers and the digital age has ushered in a new era of connectivity and information sharing. We explore how the rise of the internet, social media, and smartphones has transformed the way we communicate, access knowledge, and interact with the world around us. The digital revolution has bridged geographical gaps, creating a globally interconnected society. The digital age brought with it a seismic shift in how we interact with information and each other. The advent of the internet has turned the world into a digital village, connecting people across continents and cultures. Social media platforms have reshaped communication, enabling instant sharing of thoughts, experiences, and information. Smartphones have become extensions of ourselves, providing access to an ever-expanding digital universe. The digital revolution has given rise to virtual communities, online activism, and new forms of artistic expression, altering the way we perceive reality and define our identities.

The Rise of Artificial Intelligence (AI)

We explore the potential of artificial intelligence (AI) and machine learning to revolutionize industries and human endeavors. From self-driving cars to medical diagnostics, AI's ability to process and analyze vast amounts of data has opened new frontiers of possibility. However, we also dig into concerns about AI's impact on employment, decision-making, and the ethical considerations surrounding autonomous systems.

Artificial intelligence is no longer the realm of science fiction; it's a tangible force shaping industries and economies. Machine learning algorithms can analyze vast datasets, making predictions and decisions with remarkable accuracy. AI-powered applications span medical diagnostics, financial analysis, and even creative pursuits like music composition. Yet, as AI advances, ethical concerns emerge—questions of algorithmic bias, transparency, and accountability that demand careful consideration.

AI, AUTOMATION, AND THE FUTURE OF WORK AND ETHICS

Automation has the potential to streamline processes and increase efficiency, but it also raises questions about the future of human employment. We examine the impact of automation on various industries, from manufacturing to customer service, and explore strategies for ensuring a harmonious transition for the workforce. As technology advances, we stand at the cusp of a revolution that could redefine the nature of work and human capabilities. This section delves into the rise of artificial intelligence, automation, and the ethical considerations that accompany these transformative shifts.

Automation is redefining the landscape of employment, impacting industries from manufacturing to customer service. Robots and algorithms can execute repetitive tasks with precision and speed, enhancing efficiency and reducing error rates. However, the integration of automation also raises concerns about job displacement and the potential need for reskilling and upskilling the work- force. Society must navigate the transition to a hybrid workforce, where humans collaborate with machines in synergistic harmony.

Navigating the Transition to a Hybrid Workforce: Humans and Machines in Synergistic Harmony

1. Reskilling and Upskilling:

Invest in robust reskilling and upskilling programs to prepare the workforce for new roles that complement automation.

Training should focus on developing skills that are uniquely human, such as creativity, critical thinking, emotional intelligence, and complex problem-solving.

Investing in reskilling and upskilling programs is essential to ensure that the workforce is equipped with the necessary skills to complement automation. For instance, a manufacturing company implementing AI-driven assembly robots can offer training programs for employees to learn how to oversee and maintain these machines, as well as develop skills in quality control and problem-solving.

2. Redefining Roles to Leverage Human Strengths in the Era of AI

Rethink job roles and responsibilities to leverage the strengths of both humans and machines. Design roles that capitalize on human creativity and empathy while letting machines handle routine tasks.

In the rapidly evolving landscape of AI, redefining roles

to capitalize on human strengths is crucial for optimizing the symbiotic relationship between humans and machines. Here's an in-depth exploration of how roles can be redefined in customer service as an example:

Traditional Customer Service: Traditionally, customer service roles involved addressing a wide range of customer inquiries, from simple requests to complex issues. This often led to a high volume of routine queries being handled by human representatives, which could be time-consuming and repetitive.

REDEFINING CUSTOMER SERVICE ROLES

AI-Powered Chatbots for Routine Inquiries: By implementing AI-powered chatbots, routine inquiries can be efficiently managed. Chatbots excel at handling repetitive tasks, providing instant responses, and retrieving information from databases.

Example: When a customer contacts a company to track an order's status or inquire about store hours, an AI chatbot can quickly provide accurate information.

HUMAN REPRESENTATIVES FOR COMPLEX ISSUES

Human representatives can focus on addressing complex issues that require emotional intelligence, empathy, and critical thinking. These are qualities that AI lacks and that human interactions thrive upon.

Example: For sensitive issues like resolving billing discrepancies or addressing customer complaints, human representatives can lend a compassionate ear, negotiate solutions, and ensure the customer feels valued.

BUILDING LASTING CUSTOMER RELATIONSHIPS

Human representatives can play a pivotal role in building rapport and long-term relationships with customers. They can provide personalized recommendations, engage in active listening, and offer tailored solutions that take into account the customer's history and preferences.

Example: Human representatives in luxury retail can provide personalized fashion advice, suggesting outfits based on the customer's style preferences and previous purchases.

CREATIVE PROBLEM-SOLVING AND UPSELLING

Human representatives can engage in creative problem-solving and upselling by identifying opportunities to enhance the customer experience.

Example: In the hospitality industry, if a hotel guest encounters an issue during their stay, a human representative can not only resolve the problem promptly but also offer complimentary amenities to enhance their overall experience.

Emotional Connection and Empathy

Human representatives can establish emotional connections that are difficult for AI to replicate. They can empathize with customers' emotions, provide reassurance, and offer a human touch that fosters loyalty.

Example: When a customer contacts a telecommunications company due to service disruptions, a human representative can convey empathy and assure them that the issue is being addressed promptly.

Benefits of Redefining Roles

Efficiency: Routine queries are resolved quickly and accurately by AI, freeing up human representatives for more complex tasks.

Personalization: Human representatives can provide personalized and empathetic solutions that resonate with individual customers.

Customer Satisfaction: Customers receive prompt responses and high-quality assistance, leading to improved satisfaction.

Employee Satisfaction: Human representatives feel more engaged as they can focus on meaningful interactions.

Efficient Resource Allocation: Companies can optimize human resources for tasks that truly require human intervention.

AI is not here to take away jobs from humans, on the contrary it is here to assist humans in ways that allow for more to be done in a relatively short amount of time. Redefining roles in the era of AI involves strategically assigning tasks to the most suitable entity—whether AI or humans—based on their strengths. In customer service, this means harnessing AI's efficiency for routine inquiries while allowing human representatives to excel in emotional connection, empathy, and creative problem-solving. By achieving this balance, companies can elevate customer experiences, increase employee satisfaction, and position themselves for success in the AI-driven landscape.

Collaborative Workflows

Foster collaboration between humans and machines by designing workflows that maximize the strengths of each. This can involve humans overseeing, guiding, and enhancing automated processes. Creating workflows that maximize human-machine collaboration can significantly enhance efficiency.

Example: In medical diagnostics, radiologists can collaborate with AI algorithms to analyze medical images. The AI identifies patterns, while the radiologist provides clinical context and expertise.

Ethical Considerations

Address ethical questions arising from automation, including accountability for errors made by machines and the potential for bias in AI algorithms.

Addressing ethical concerns includes ensuring AI systems are transparent, accountable, and unbiased.

Example: Financial institutions using AI for loan approvals must monitor for biases in algorithmic decisions and have mechanisms in place to correct biases if identified. The Human Role in AI-Powered Loan Approvals and Bias Mitigation is a critical one. In financial institutions that utilize AI for loan approvals, humans play a pivotal role in ensuring fair and unbiased decisions while leveraging the efficiency of AI algorithms. Here's an in-depth exploration of their roles and how they monitor and mitigate biases:

Human Oversight and Decision-Making: Human experts are responsible for designing, training, and maintaining AI algorithms used for loan approvals. They define the parameters and criteria that guide the algorithms' decisions, ensuring that they align with legal, ethical, and business requirements.

Training Data Selection: Humans curate the training datasets used to teach AI models. They ensure that the data is diverse, representative, and free from discriminatory patterns. This step is crucial to prevent AI algorithms from learning biased patterns present in historical data.

Bias Detection and Mitigation: Humans actively monitor AI-generated loan decisions to detect any biases that may arise from the algorithm's decision-making process.

Regular Audits: Experts regularly audit the AI algorithms to identify instances where certain demographic groups might be disproportionately affected by the decisions. They examine statistical measures and trends to spot potential disparities.

Developing Fairness Metrics: Humans design fairness metrics to assess algorithmic decisions across different demographic groups. These metrics provide quantifiable measures of fairness and help identify disparities that need correction.

Bias Correction Strategies: If biases are detected, human experts collaborate with data scientists to adjust the algorithms' parameters and training data. This might involve reweighting the training data, fine-tuning decision thresh- olds, or modifying the model's architecture.

Transparency and Change to Explainability: Experts ensure that the decision-making process of AI algorithms is transparent and explainable. This involves developing methods to provide insights into how the AI arrives at its decisions, enabling accountability.

Continuous Learning: Human experts stay informed about advancements in AI fairness research. They learn from the experiences of other institutions and incorporate best practices into their own processes.

Feedback Loop: Input from affected parties, customers, and advocacy groups is invaluable in identifying potential biases and understanding their real-world impact. Human experts encourage an ongoing feedback loop to improve algorithmic fairness.

Legal and Ethical Compliance: Human experts ensure that the AI algorithms adhere to legal requirements, such as anti-discrimination laws, and follow ethical guidelines set by the organization and industry.

Bias Testing Tools: Experts use specialized tools to simulate and test how AI algorithms respond to different inputs. These tools help identify hidden biases that might not be immediately evident.

Diversity in Teams: Maintaining diverse teams that include experts from various backgrounds ensures a more comprehensive and balanced perspective on bias detection and mitigation.

BENEFITS OF HUMAN OVERSIGHT

Ethical AI Deployment: Human experts ensure that AI algorithms are deployed ethically, minimizing harm and bias to individuals and groups.

Enhanced Accountability: Humans provide an accountable and transparent decision-making process, building trust among customers and stakeholders.

Flexibility in Decision-Making: Experts can apply contextual understanding to unique cases that may not conform to automated rules.

Adapting to Evolving Context: Humans can adjust algorithms to respond to shifting societal norms and new challenges.

The role of humans in financial institutions that use AI for loan approvals is indispensable for ensuring fair, transparent, and unbiased decisions. By actively monitoring for biases, correcting algorithmic behavior, and maintaining transparency, these experts ensure that AI algorithms align with legal and ethical standards. Through collaboration between humans and AI, financial institutions can harness the efficiency of technology while upholding fairness and equity in lending decisions.

Flexibility and Adaptability

Cultivate a culture of continuous learning and adaptability, enabling employees to navigate changes in job requirements and work alongside machines. Cultivating adaptability helps employees navigate changes in job requirements. Example: Retail workers might transition from manual inventory management to using automated inventory systems and simultaneously develop skills in customer engagement and relationship-building.

Human-Centered Design

Prioritize human-centered design in automation deployment, ensuring that technology serves to enhance human well-being, productivity, and job satisfaction. Prioritizing human well-being ensures that AI technologies enhance work experiences.

Example: Designing AI-driven scheduling tools that consider employees' preferences and workloads, ultimately leading to better work-life balance.

INCLUSIVE DECISION-MAKING

Involve employees in decisions related to automation implementation to ensure that their insights and concerns are considered. Involving employees in automation decisions ensures diverse perspectives.

Example: Manufacturing workers contributing insights on which processes can be automated, is a way of ensuring that human expertise is maintained where needed.

HYBRID ROLES

Create hybrid roles that combine technical skills with interpersonal skills. For example, a role might involve both data analysis and communication with stakeholders. Creating hybrid roles merges technical skills with interpersonal abilities.

Example: Data analysts not only crunch numbers but also work closely with marketing teams, translating data insights into effective communication strategies.

Redistributing Workloads

Use automation to distribute workloads more evenly, reducing burnout and freeing up time for more value-added tasks. Automation can redistribute workloads, reducing repetitive tasks for humans.

Example: HR professionals can use AI for initial candidate screening, allowing them to focus more on interviewing and building strong relationships with potential hires.

Lifelong Learning

Promote lifelong learning by offering opportunities for employees to acquire new skills and knowledge throughout their careers. Promoting lifelong learning encourages continuous skill development.

Example: A financial institution offers employees opportunities to learn new coding languages, enabling them to collaborate with AI developers on designing financial models.

Job Design and Satisfaction

Redesign jobs to focus on tasks that are engaging and fulfilling for employees, allowing them to contribute meaningfully to their organizations. Redesigned jobs focus on fulfilling tasks.

Example: In manufacturing, roles can shift from

monotonous assembly line work to supervising and improving automated processes, leading to higher job satisfaction.

Cross-Training

Cross-train employees to have a broader skill set, enabling them to switch between tasks that require different levels of automation. Cross-training equips employees for diverse tasks.

Example: A customer service representative trained to handle both customer interactions and manage the AI chatbot system.

Human-Machine Interaction

Develop interfaces and interactions that facilitate seamless collaboration between humans and machines, enhancing overall efficiency. Seamless interaction between humans and machines enhances efficiency.

Example: Engineers working with AI-driven design tools that optimize parameters based on human inputs and machine-generated suggestions.

Social Safety Nets

Implement social safety nets to support workers who face job displacement due to automation, ensuring a just transition. While AI is here to lend a helping hand so humans can accomplish more tasks in relatively short periods, it may be difficult for some to keep up with the changes that come with it. Safety nets therefore protect workers facing job displacement.

Example: Governments implementing income support programs for workers affected by automation-induced job losses.

INCLUSIVE GROWTH

Use the benefits of automation to drive inclusive economic growth, creating opportunities for underrepresented groups and marginalized communities. Using automation benefits to drive inclusive growth ensures broader societal advancement.

Example: A tech company using a portion of its automation-derived profits to fund STEM education programs in underserved communities.

Navigating the transition to a hybrid workforce requires a holistic approach that takes into account the evolving nature of work, the potential of automation, and the well- being of the workforce. By strategically combining human creativity, adaptability, and emotional intelligence with the precision and speed of machines, societies can shape a future where both humans and machines thrive in synergistic harmony. Navigating the transition to a hybrid workforce requires a multifaceted approach that involves upskilling, role redesign, collaborative workflows, and ethical considerations. The goal is to leverage the unique strengths of both humans and machines to create a harmonious and efficient work environment that contributes to personal growth, organizational success, and societal progress.

Ethical Considerations: Privacy, Security, and Bias

The proliferation of technology has raised concerns about data privacy, cybersecurity, and algorithmic bias. We delve into the ethical dilemmas surrounding the collection and use of personal data, the vulnerabilities of digital systems, and the need to ensure fairness and inclusivity in technology design.

Environmental Impact and Sustainability

Technological progress has brought undeniable benefits, but it has also placed strain on the environment. We explore the environmental implications of resource extraction, energy consumption, and electronic waste. Examining sustainable design, renewable energy, and circular economy models, we consider how technology can contribute to a more environmentally conscious future.

Cultivating Digital Literacy and Responsible Citizenship

In a world driven by technology, digital literacy and responsible citizenship become essential. We discuss the importance of fostering critical thinking skills, media literacy, and ethical decision-making in an age of information overload and digital manipulation.

In the realm of technology, humanity stands at a crossroads of limitless potential and complex challenges. As we navigate the impact of technological advancements on society, harness the capabilities of AI and automation, and confront the ethical considerations that arise in a technology-driven world, we are called to balance progress with responsibility. Stay tuned as we journey through the intricate landscape of innovation, contemplating the boundless horizons that lie ahead while charting a course of ethical stewardship for the technology that shapes our lives.

ADDRESSING ETHICAL CONCERNS IN AI

- Algorithmic Bias:

Efforts are being made to identify and rectify biases in AI algorithms, particularly in areas like criminal justice and hiring, where biased decisions can have serious consequences.

Developing more diverse and inclusive teams of AI researchers and practitioners can help reduce biases in algorithm design.

Regular audits and testing of AI systems for bias are crucial to ensure fairness and equity.

- Transparency and Accountability:

Organizations are working on creating transparent AI systems that provide explanations for their decisions.

Regulatory frameworks are being developed to ensure that AI developers are held accountable for their creations, with provisions for transparency and accountability in AI applications.

- Data Privacy and Security:

Stricter data protection regulations, such as GDPR, are in place to safeguard personal data used by AI systems.

Advances in techniques like federated learning allow AI models to be trained on decentralized data while maintaining privacy.

- Robustness and Reliability:

Researchers are developing methods to make AI systems more robust against adversarial attacks and unexpected inputs.

Continuous testing and validation of AI models are necessary to ensure they perform reliably in real-world scenarios.

- Collaboration and Standards:

Collaboration among governments, industries, and academia is essential to establish common ethical standards and guidelines for AI development and deployment.

Organizations like Institute of Electrical and Electronics Engineering (IEEE) and OpenAI are working to create frameworks and standards for responsible AI development.

Consequences of Not Addressing Ethical Concerns

- Reinforcing Bias and Inequality: If not addressed, algorithmic biases can perpetuate and even exacerbate existing social biases, leading to unfair treatment and discrimination.

- Lack of Trust: Without transparency and accountability, users and stakeholders may lose trust in AI systems, hindering their adoption in critical areas.

Legal and Regulatory Issues

Failure to address ethical concerns can lead to legal challenges, fines, and reputational damage for organizations deploying biased or unethical AI systems.

Safety Risks

AI systems with inadequate robustness can lead to unintended consequences, such as autonomous vehicles making dangerous decisions in unpredictable situations.

SOCIAL UNREST

If AI contributes to job displacement without adequate measures for retraining and transition, it can lead to economic disparities and social unrest.

MISSED OPPORTUNITIES

Ignoring ethical concerns could deter investment in AI technologies, hindering the potential benefits they could bring to industries and economies.

Addressing ethical concerns in AI is not only a moral imperative but also a practical necessity for the responsible development and deployment of AI technologies. By proactively working to mitigate biases, ensuring transparency, upholding accountability, and fostering collaboration, we can harness the potential of AI while minimizing its potential negative impacts. Failing to address these concerns could lead to dire consequences that not only hinder progress but also have wide-ranging societal implications.

Chapter 7

Unlocking the Human Mind

The enigma of consciousness and self-awareness.

The human mind is a realm of boundless complexity, a theater where thoughts, emotions, and perceptions dance in intricate patterns. In this chapter, we venture into the enigma of consciousness and self-awareness, including the philosophical and scientific mysteries that surround the inner workings of the human mind. The human mind, with its mysteries and complexities, remains a frontier of exploration, inviting us to journey inward and unravel the profound enigmas that define our existence. As we venture into the depths of consciousness, traverse the neural landscapes of the brain, and contemplate altered states of awareness, we gain new insights into the nature of our minds and the intricate tapestry of human cognition. Read on as we unlock the doors to the mind's labyrinth, where each step brings us closer to understanding the essence of who we are and the wonders that lie within.

The Nature of Consciousness

Consciousness—the essence of subjective experience—remains one of the most profound puzzles in human understanding. We explore philosophical inquiries into the nature of consciousness, from dualism to materialism, and the elusive question of how the brain gives rise to our conscious awareness.

EXPLORING THE PHILOSOPHICAL NATURE OF CONSCIOUSNESS: DUALISM TO MATERIALISM

The nature of consciousness is a philosophical puzzle that has fascinated thinkers for centuries. How do subjective experiences, thoughts, and self-awareness arise from the physical processes of the brain? This intricate inquiry has given rise to several philosophical perspectives, each attempting to unravel the enigma of consciousness.

DUALISM: Dualism posits that consciousness, and the physical body are two distinct entities. This perspective has ancient roots, with notable proponents like René Descartes. Cartesian dualism, for instance, suggests that the mind and body interact through a non-physical interface. The mind is immaterial and possesses self-awareness, while the body functions in the material world. This approach raises challenging questions about how these separate realms interact and influence each other.

IDEALISM: Idealism, closely related to dualism, asserts that reality is fundamentally mental or immaterial. This philosophy posits that consciousness is primary, and the external world arises from our perceptions and thoughts. While idealism offers an intriguing perspective on the nature of reality, it faces skepticism regarding the relationship between individual consciousness and the shared external world.

MATERIALISM: Materialism asserts that consciousness is an emergent property of the physical brain. This perspective gained prominence with the rise of modern science and the belief that all phenomena, including consciousness, can ultimately be explained through the laws of physics and chemistry. Materialism aligns closely with the scientific method, aiming to understand consciousness as a product of complex neural processes.

The Elusive Question of Conscious Awareness and the Brain

The puzzle of consciousness becomes particularly intricate when we examine how the brain, a physical organ composed of neurons and synapses, gives rise to the phenomenon of conscious awareness. This question, often referred to as the "hard problem of consciousness," delves into the essence of what it means to be conscious and how this experience emerges from neural activity.

NEURAL CORRELATES OF CONSCIOUSNESS: Neuroscience has made significant strides in identifying neural correlates of consciousness—the brain regions and processes that

seem to be closely associated with conscious experiences. Brain imaging techniques, such as functional MRI, have revealed patterns of brain activity that coincide with various mental states and sensory perceptions. However, identifying these neural correlates doesn't fully explain the nature of conscious awareness.

BINDING PROBLEM AND UNITY OF CONSCIOUSNESS: A central challenge in understanding consciousness is the "binding problem." Our conscious experiences are unified and coherent, yet the brain processes information through various specialized areas. How does the brain integrate these fragmented signals into a seamless conscious experience? The binding problem highlights the intricacy of how different aspects of consciousness, such as perception, emotion, and self-awareness, come together.

EMERGENCE AND COMPLEXITY: Materialist views propose that consciousness emerges from complex interactions among neurons, synapses, and neural networks. This emergence is akin to how water's properties emerge from the interaction of hydrogen and oxygen atoms. However, the question remains: How does the aggregation of physical processes result in subjective experiences and self- awareness?

THE PHILOSOPHICAL CHALLENGE: The challenge of explaining how the brain generates consciousness is not merely a scientific inquiry; it's a philosophical dilemma that delves into the nature of reality, the mind-body relationship, and the boundaries of human understanding. The gap between physical processes and conscious experiences poses a profound question that stretches the limits of our current knowledge.

In the exploration of consciousness—from dualism's separation of mind and body to materialism's focus on neural processes—we grapple with fundamental questions about the essence of our being. The elusive nature of conscious awareness challenges our understanding of the physical and metaphysical aspects of reality, inviting us to ponder the ultimate connection between the material brain and the immaterial realm of consciousness.

Self-Awareness and Identity

At the heart of human consciousness lies self-awareness— the ability to introspect and recognize oneself as an individual entity. We investigate how self-awareness emerges in childhood, its role in shaping our identities, and the intricate interplay between our perceptions, memories, and sense of self.

EXPLORING THE EMERGENCE OF SELF-AWARENESS IN CHILDHOOD

Self-awareness, the ability to recognize oneself as a distinct individual, is a remarkable cognitive achievement that unfolds in the early stages of human development.

This journey from mere perception to self-recognition is a complex process that shapes our understanding of ourselves and our place in the world.

MIRROR SELF-RECOGNITION: One of the most recognized milestones in the development of self-awareness is mirror self-recognition. Around the age of 18 to 24 months, children begin to recognize themselves in a mirror, indicating an awareness of their reflection as separate from others. This ability signifies a cognitive shift from perceiving oneself as part of the environment to recognizing oneself as a distinct entity.

ROLE OF SOCIAL INTERACTION: Social interactions play a pivotal role in the emergence of self-awareness. Caregivers, through interactions and feedback, help infants form a sense of agency and distinction. By responding to the child's cues and actions, caregivers validate the child's existence as an individual with intentions and desires. Over time, these interactions contribute to the development of self-awareness.

DEVELOPING A THEORY OF MIND: Self-awareness is intricately linked to the development of a "theory of mind," which is the ability to attribute mental states to oneself and others. As children acquire the understanding that others have thoughts, emotions, and intentions separate from their own, they also develop a deeper understanding of their own internal experiences.

ROLE OF LANGUAGE: Language acts as a tool for self-expression and introspection. As children acquire language skills, they gain the ability to communicate their thoughts, desires, and emotions, leading to greater self-awareness. Language facilitates the internalization of thoughts, allowing children to reflect on their experiences and construct narratives about themselves.

SHAPING OUR IDENTITIES: Self-awareness forms the bedrock of identity—the multifaceted sense of who we are as individuals. It influences how we perceive ourselves, our roles in relationships, and our place in society.

SELF-CONCEPT: Self-awareness contributes to the development of a self-concept—a mental representation of one's traits, characteristics, and values. This self-concept evolves as individuals gain a deeper understanding of their abilities, interests, and social roles. It acts as a lens through which we interpret the world and make sense of our experiences.

SOCIAL IDENTITY: As self-awareness matures, individuals begin to form a sense of social identity, which is linked to group memberships such as family, culture, and nationality. Social identity shapes our beliefs, values, and behaviors and influences how we perceive others and ourselves in relation to these groups.

THE INTERPLAY BETWEEN PERCEPTIONS, MEMORIES, AND SENSE OF SELF: The formation of self-awareness involves a complex interplay between our perceptions, memories, and sense of self. These elements interact to create a cohesive and evolving sense of who we are.

PERCEPTIONS AND SELF-IMAGE: Our perceptions of ourselves, including how we look, how we interact with others, and how others react to us, shape our self-image. Positive or negative perceptions can significantly influence our self-esteem and confidence.

MEMORIES AND NARRATIVE IDENTITY: Memories contribute to the construction of our narrative identity—a cohesive story of our past, present, and future. As we recall and interpret past experiences, we weave a narrative that provides continuity and meaning to our lives.

SENSE OF SELF AS DYNAMIC AND FLUID: The sense of self is not static; it evolves in response to new experiences, challenges, and personal growth. Our interactions with the world and our reflections continuously shape our sense of identity.

In the intricate dance of self-awareness, we witness the emergence of individuality, the shaping of identities, and the interplay between our perceptions, memories, and sense of self. From early childhood's first recognition of one self in a mirror to the intricate web of social interactions, language, and introspection, the journey of self-awareness is a profound exploration of who we are and how we relate to the world around us.

Neuroscience and the mysteries of the human brain

Advances in neuroscience have unraveled many mysteries of the human brain, yet its complexity remains unparalleled. This section takes us on a journey through the neural labyrinth, exploring the intricate architecture and functions of the brain.

JOURNEYING THROUGH THE NEURAL LABYRINTH: Unraveling the Intricate Architecture and Functions of the Brain

The human brain stands as the pinnacle of biological complexity, a three-pound organ that holds within it the essence of our thoughts, emotions, and consciousness. While modern neuroscience has made remarkable strides in deciphering its mysteries, the brain's intricate architecture and functions continue to captivate and challenge our understanding.

The Brain's Architectural Marvel

NEURONS AND SYNAPSES: At the core of the brain's complexity are neurons, specialized cells that transmit information through electrical and chemical signals. These neurons form intricate networks, connecting via synapses— tiny gaps where information is transmitted from one neuron to another. The sheer number of neurons and synapses in the brain is staggering, with estimates reaching trillions.

CEREBRAL CORTEX: The cerebral cortex, often referred to as

the brain's "gray matter," is responsible for higher-order cognitive functions. It's divided into lobes, each associated with specific functions such as sensory perception, motor control, language processing, and executive functions like decision-making and problem-solving.

SUBCORTICAL STRUCTURES: Beneath the cerebral cortex lie subcortical structures that play critical roles in functions like emotion regulation, memory formation, and the regulation of vital bodily processes. The amygdala, for instance, is involved in processing emotions, while the hippocampus is crucial for memory consolidation.

FUNCTIONS OF THE BRAIN:

SENSORY PROCESSING: The brain's sensory regions receive and process information from our senses—sight, sound, touch, taste, and smell. Visual processing occurs in the occipital lobe, auditory processing in the temporal lobe, and somatosensory processing in the parietal lobe.

MOTOR CONTROL: The brain's motor cortex and associated regions control our voluntary movements. The intricate coordination required for actions as simple as reaching for an object involves a symphony of neural signals and feedback loops.

LANGUAGE AND COMMUNICATION: Language processing is distributed across various brain regions, including Broca's area, responsible for speech production, and Wernicke's area, responsible for language comprehension. The brain's ability to seamlessly comprehend and produce language is a testament to its complexity.

EMOTION AND MEMORY: The limbic system, including structures like the amygdala and hippocampus, plays a pivotal role in processing emotions and forming memories. Emotional experiences are deeply intertwined with memory formation and retrieval.

EXECUTIVE FUNCTIONS: The prefrontal cortex, often considered the brain's command center, is responsible for executive functions such as planning, decision-making, impulse control, and working memory. These functions are critical for goal-directed behavior and adapting to complex environments.

NETWORK CONNECTIVITY: Beyond individual regions, the brain operates as a network of interconnected areas. Neural pathways allow for the exchange of information across different brain regions, facilitating complex cognitive processes.

CHALLENGES AND UNANSWERED QUESTIONS:

CONSCIOUSNESS AND THE BINDING PROBLEM: While we've identified neural correlates of consciousness, the question of how disparate brain regions bind together to create our unified conscious experience—the "binding problem"—remains a fundamental challenge.

NEUROPLASTICITY AND LEARNING: Neuroplasticity, the brain's

ability to reorganize itself in response to experience, underpins our ability to learn and adapt. However, the mechanisms that drive neuroplasticity and the extent of its potential are still being explored.

BRAIN DISORDERS AND MENTAL HEALTH: Despite advancements, our understanding of brain disorders such as Alzheimer's, schizophrenia, and depression remains incomplete. Unraveling the neural underpinnings of these conditions is crucial for developing effective treatments.

As we journey through the neural labyrinth, we encounter a world of awe-inspiring complexity. The brain's architecture and functions hold the key to understanding the essence of human cognition, emotion, and consciousness. Yet, with each discovery, new questions arise, reminding us that the exploration of the brain's mysteries is a lifelong endeavor—a voyage that continues to unfold with every neuron fired and synapse formed.

Mapping the Brain: Neuroimaging and Connectivity:

Modern neuroscience employs a range of tools, from MRI scans to brain-mapping techniques, to illuminate the brain's structure and connectivity. We explore the regions responsible for cognition, emotion, and motor control, and how their interactions give rise to our thoughts and behaviors.

Exploring the Dynamic Interplay of Brain Regions: Cognition, Emotion, and Motor Control

The human brain is a masterpiece of organization, with distinct regions responsible for a myriad of functions. These regions seamlessly interact, creating a rich tapestry of our thoughts, emotions, and behaviors. Let's embark on a journey through the brain to understand how the regions responsible for cognition, emotion, and motor control collaborate to shape our complex human experience.

Cognition: The Cognitive Network

Prefrontal Cortex: At the forefront of cognition lies the prefrontal cortex. This region, situated at the front of the brain, is the seat of executive functions—higher-order cognitive processes that guide decision-making, planning, and problem-solving. The prefrontal cortex is divided into subregions, each contributing to specific aspects of cognitive control and social behavior.

Parietal Lobe: The parietal lobe processes sensory information and plays a crucial role in spatial awareness, attention, and perception. It integrates sensory inputs and helps us navigate our environment, enabling tasks like judging distances, recognizing objects, and understanding spatial relationships.

Temporal Lobe: Nestled beneath the lateral fissure, the temporal lobe is central to memory, language processing, and auditory perception. The hippocampus, within the temporal lobe, is vital for forming new memories and spatial navigation.

EMOTION: THE EMOTIONAL NETWORK

AMYGDALA: The amygdala, an almond-shaped structure, isa hub for processing emotions. It plays a central role in recognizing and interpreting emotional cues, especially those related to fear and threat. The amygdala's interactions with other regions help to regulate emotional responses and guide decision-making based on emotional significance.

CINGULATE CORTEX: Situated in the medial part of the brain, the cingulate cortex is involved in regulating emotions, pain perception, and decision-making. It helps monitor conflicts between competing thoughts and emotions, guiding our responses to emotional stimuli.

HYPOTHALAMUS: Deep within the brain, the hypothalamus orchestrates a range of bodily functions, including the autonomic nervous system, hormone regulation, and the expression of emotions. It connects the brain to the body's physiological responses to emotions like stress, pleasure, and fear.

MOTOR CONTROL: THE MOTOR NETWORK

MOTOR CORTEX: The motor cortex, located in the frontal lobe, is the command center for voluntary movements. It sends signals to muscles, enabling us to perform intricate actions—from delicate finger movements to complex athletic feats. Different parts of the motor cortex are associated with specific body areas, forming a "motor homunculus."

CEREBELLUM: Often referred to as the "little brain," the cerebellum is responsible for coordinating movement, balance, and posture. It fine-tunes motor commands and helps maintain fluid movements, contributing to our ability to walk, run, and engage in skilled activities.

BASAL GANGLIA: The basal ganglia play a crucial role in controlling voluntary movements and habits. They help initiate and regulate actions, allowing for smooth coordination between different muscle groups.

Interplay and Integration: Thoughts and Behaviors

The interactions between these regions are intricate and dynamic, creating the landscape of our thoughts and behaviors.

COGNITION-EMOTION INTERACTION: The interplay between cognitive and emotional regions is evident in decision-making. The prefrontal cortex assesses options, while the emotional regions, like the amygdala, contribute emotional valence. This interaction guides choices that align with our values, goals, and emotional well-being.

MOTOR-COGNITION CONNECTION: The motor regions connect with cognitive regions, enabling us to translate thoughts into actions. The prefrontal cortex plans, the motor cortex executes, and feedback loops allow us to adjust movements based on sensory input.

EMOTION-MOTORINTEGRATION: Emotional responses often translate into motor actions. Fear may trigger the "fight or flight" response, while happiness might lead to joyful expressions. These responses are orchestrated by the emotional regions' connections to motor control areas.

In the mosaic of the brain, the regions responsible for cognition, emotion, and motor control are intertwined, weaving a complex symphony that guides our thoughts, feelings, and actions. As we navigate the intricate connections and interplay between these regions, we gain insight into the mechanisms that underpin human behavior and the remarkable ways in which our brains orchestrate our experiences in the world.

PLASTICITY AND LEARNING:

The brain's capacity for change and adaptation, known as neuroplasticity, underpins our ability to learn and acquire new skills throughout life. We investigate the mechanisms of neuroplasticity and how experiences and learning reshape the brain's connections, influencing our cognitive development.

UNVEILING THE MARVEL OF NEUROPLASTICITY: Reshaping the Brain's Connections Through Experience and Learning

Neuroplasticity, often referred to as the brain's "plasticity" or "neural plasticity," is a cornerstone of human cognition. This remarkable phenomenon allows the brain to adapt, rewire, and reshape its connections in response to experiences and learning. Let's embark on a journey into the intricate mechanisms of neuroplasticity and understand how this phenomenon underpins our cognitive development.

MECHANISMS OF NEUROPLASTICITY:

SYNAPTIC PLASTICITY: At the heart of neuroplasticity lies synaptic plasticity—the ability of synapses, the junctions between neurons, to strengthen or weaken based on neural activity. Long-Term Potentiation (LTP) and Long-Term Depression (LTD) are processes through which synapses undergo lasting changes in their efficiency, strengthening or weakening the connection between neurons.

STRUCTURAL PLASTICITY: Structural plasticity involves changes in the physical structure of neurons and their connections. This includes the growth of new dendritic branches, the formation of new synapses, and even the elimination of existing ones through a process known as synaptic pruning.

EXPERIENCE-DEPENDENT NEUROPLASTICITY:

LEARNING AND MEMORY: Learning is a prime driver of neuroplasticity. When we engage in new experiences, acquire knowledge, or practice skills, our brain's networks reorganize to accommodate this information. Memory formation involves strengthening the connections among neurons involved in encoding and retrieving specific memories.

HEBBIAN LEARNING: The famous "neurons that fire together, wire together" adage captures the essence of Hebbian learning—a foundational principle of experience-dependent plasticity. When two neurons are activated simultaneously, their synaptic connection strengthens, fostering efficient communication.

USE-DEPENDENT PLASTICITY: Neural circuits become more robust with use and weakened with disuse—a phenomenon known as use-dependent plasticity. For instance, musicians' brain regions responsible for motor control and auditory processing may undergo structural changes as they continually practice and refine their skills.

SYNAPTIC SCALING AND HOMEOSTATIC PLASTICITY:

SYNAPTIC SCALING: To prevent neural circuits from becoming overly excitable or inactive, the brain employs synaptic scaling—a mechanism that adjusts the strength of synapses in response to overall network activity. This maintains a delicate balance in the firing rates of neurons.

HOMEOSTATIC PLASTICITY: Homeostatic plasticity ensures that neuronal networks remain stable despite constant changes. If a network becomes too active, homeostatic mechanisms weaken synapses; conversely, if a network becomes less active, synapses may strengthen.

THE BRAIN'S JOURNEY OF ADAPTATION:

SKILL ACQUISITION AND EXPERTISE: Learning new skills involves neural circuits working in concert. As we practice, specific pathways strengthen, leading to improved performance. Over time, with continued practice, we transition from conscious effort to automaticity as neural pathways become finely tuned.

RECOVERY FROM BRAIN INJURY: Neuroplasticity is a vital player in brain recovery after injury. In response to damage, neighboring regions can take on the functions of the affected area—a phenomenon known as cortical remapping.

ENVIRONMENTAL ENRICHMENT: The brain thrives in a stimulating environment. Enriched environments with diverse sensory experiences can stimulate the growth of new neurons and enhance synaptic connectivity.

INFLUENCES ON COGNITIVE DEVELOPMENT:

Neuroplasticity is a lifelong journey. It shapes cognitive development across the lifespan, from infancy through old age. Early experiences, such as language exposure in childhood, sculpt neural networks that support language processing. However, the brain's plasticity doesn't wane in adulthood—lifelong learning, mental exercises, and novel experiences continue to drive adaptation.

In the intricate dance of experience and learning, neuroplasticity orchestrates the symphony of our cognitive development. Whether we're mastering a musical instrument, navigating a new language, or adapting to challenges, our brain's ability to rewire and reshape itself stands as a testament to its remarkable adaptability. As we embrace the notion that our brains are constantly evolving, we open the door to a lifelong journey of growth and discovery.

Top of Form

EXPLORING ALTERED STATES OF CONSCIOUSNESS AND THEIR IMPLICATIONS.

Human consciousness is not fixed; it can transcend ordinary states through meditation, dreams, or even substances. This section digs into altered states of consciousness and their implications for our understanding of the mind.

MEDITATION AND ALTERED STATES:

Meditative practices have been instrumental in inducing altered states of consciousness, expanding awareness beyond the confines of everyday thought. We explore the scientific exploration of meditation's effects on the brain, as well as its potential for enhancing well-being, focus, and emotional regulation.

THE SCIENCE OF MEDITATION: Unveiling the Effects on the Brain and Enhancing Well-being

Meditation, a practice that has ancient roots in contemplative traditions, has become a subject of rigorous scientific investigation in recent decades. The exploration of meditation's effects on the brain has revealed remarkable insights into its potential for enhancing well-being, improving focus, and regulating emotions. Let's get into the scientific journey that has illuminated the transformative power of meditation.

MEDITATION AND BRAIN PLASTICITY:

STRUCTURAL CHANGES: Brain imaging studies have shown that regular meditation practice can lead to structural changes in the brain. Regions associated with attention, emotion regulation, and self-awareness, such as the prefrontal cortex and insula, exhibit increased gray matter volume in experienced meditators.

CORTICAL THICKNESS: Meditation has been linked to thicker cortical regions, suggesting enhanced neural connections. The anterior cingulate cortex, which plays a role in attention and emotion, is one such region that appears to benefit from meditation-induced plasticity.

EFFECTS ON BRAIN FUNCTION:

ATTENTION AND FOCUS: Mindfulness meditation, in particular, has been extensively studied for its effects on attention. Brain scans reveal increased activation in brain networks associated with sustained attention and reduced activation in regions related to mind-wandering, leading to improved focus and concentration.

EMOTION REGULATION: Meditation's impact on emotional regulation is evident in its ability to reshape how the brain responds to emotions. Studies show reduced reactivity in the amygdala—the brain's emotional center—coupled with increased connectivity between the prefrontal cortex and the amygdala, supporting better emotion regulation.

DEFAULT MODE NETWORK: Meditation seems to quiet the default mode network (DMN), which is active during mind-wandering and self-referential thinking. This deactivation of the DMN correlates with increased present-moment awareness and reduced rumination, contributing to emotional well-being.

NEUROCHEMICAL CHANGES:

STRESS REDUCTION: Meditation's effects extend beyond brain structure and function to neurochemical changes. Regular meditation practice has been associated with reduced levels of stress hormones like cortisol, indicating a physiological response to stressors.

SEROTONIN AND DOPAMINE: Some meditation practices have been linked to increased serotonin and dopamine production, both of which play roles in mood regulation, pleasure, and overall well-being.

MIND-BODY CONNECTION:

NEUROPLASTICITY AND BODY AWARENESS: Body-centered meditation practices, like yoga and tai chi, involve mindful movement and enhance interoceptive awareness—our ability to perceive internal bodily sensations. This connection between mind and body nurtures self-awareness and emotional regulation.

IMMUNE SYSTEM BOOST: Meditation's impact extends to the immune system. Mindfulness practices have been associated with improved immune function, possibly due to reduced inflammation and stress-related suppression.

PROMOTING WELL-BEING:

STRESS REDUCTION AND RESILIENCE: The stress-reducing effects of meditation make it a powerful tool for building resilience. By cultivating mindfulness and emotional regulation, individuals are better equipped to navigate life's challenges.

ANXIETY AND DEPRESSION: Scientific research suggests that mindfulness-based meditation interventions can be effective in reducing symptoms of anxiety and depression. Meditation helps individuals observe thoughts and emotions without judgment, disrupting negative thought patterns.

ENHANCED COGNITIVE ABILITIES: Meditation's positive impact on attention and focus extends to improved cognitive abilities. Studies show that regular practice correlates with better working memory, cognitive flexibility, and problem-solving skills.

In the journey from ancient practice to modern scientific exploration, meditation has emerged as a powerful tool for enhancing well-being, fostering focus, and regulating emotions. The brain's remarkable plasticity allows meditation to sculpt neural connections, leading to structural and functional changes that positively influence cognitive and emotional processes. As science continues to unveil the transformative potential of meditation, we are reminded of the profound connection between mind and brain—a connection that holds the promise of a more balanced, resilient, and enriched human experience.

DREAMS AND UNCONSCIOUS MIND:

Dreams offer a window into the unconscious mind, a realm of symbolism and emotion. We examine the theories of dream interpretation, exploring the role of dreams in memory consolidation, problem-solving, and emotional processing.

DREAMS: Unveiling the Unconscious Mind's Symbolism and Function

Dreams have long captivated the human imagination, offering a mysterious portal into the realm of the unconscious mind. Theories of dream interpretation have evolved over centuries, while modern research has shed light on the multifaceted functions of dreams—ranging from memory consolidation to emotional processing. Let's delve into the intricate world of dreams, their symbolism, and the roles they play in our cognitive and emotional landscapes.

THEORIES OF DREAM INTERPRETATION:

PSYCHOANALYTIC THEORY (SIGMUND FREUD): Freud's psychoanalytic theory posits that dreams are a window into repressed desires and conflicts. He believed that the symbolic content of dreams offers insights into our unconscious wishes and unprocessed emotions. Freud identified two layers of dream content: manifest content (the literal storyline) and latent content (the underlying, symbolic meaning).

ACTIVATION-SYNTHESIS THEORY (J. ALLAN HOBSON AND ROBERT MCCARLEY): The activation-synthesis theory suggests that dreams are the brain's attempt to make sense of random neural activity during REM sleep. According to this theory, the brain constructs a narrative to rationalize the chaotic signals, resulting in dream experiences that might not necessarily have deep psychological meaning.

COGNITIVE THEORY (ANTTI REVONSUO): Cognitive theories propose that dreams serve an evolutionary purpose related to threat simulation. Dreams, according to this view, allow the mind to rehearse and prepare for potential dangers, contributing to survival and problem-solving.

THE ROLE OF DREAMS IN COGNITIVE FUNCTIONS:

MEMORY CONSOLIDATION: Dreams are implicated in memory consolidation—the process by which newly acquired information is solidified and integrated into long- term memory. During REM sleep, when vivid dreaming occurs, the brain replays and reorganizes experiences from the day, enhancing memory retention.

PROBLEM-SOLVING AND CREATIVITY: Dreams have a remarkable capacity to facilitate problem-solving and foster creativity. The dream state can allow the mind to explore solutions to challenges by forming novel connections and perspectives that might not emerge in waking thought processes.

EMOTIONAL PROCESSING AND REGULATION: Dreams serve as a platform for emotional processing, allowing us to confront and process unresolved emotions. Dreams can help individuals explore fears, traumas, and unresolved conflicts in a safe, symbolic space, potentially contributing to emotional healing and growth.

DREAMS AND EMOTIONAL REGULATION:

THERAPEUTIC POTENTIAL: Dreams can be harnessed in therapeutic settings to explore and address emotional issues. Dream analysis, based on Jungian or experiential therapies, can aid in uncovering patterns, symbols, and emotions that underlie psychological struggles.

NIGHTMARES AND TRAUMA: Nightmares, distressing dreams that evoke fear or anxiety, might be the mind's way of processing traumatic experiences. Understanding and working through recurring nightmares can be therapeutic in addressing trauma-related symptoms.

LUCID DREAMING: In lucid dreaming, individuals become aware that they are dreaming and can actively participate in the dream scenario. Lucid dreaming can offer a unique opportunity to confront fears, practice skills, and explore creativity while in a controlled dream environment.

In the enigmatic realm of dreams, the unconscious mind weaves a tapestry of symbolism and emotion. While theories of dream interpretation vary, modern science unveils the multifunctional role of dreams in memory consolidation, problem-solving, and emotional processing. Dreams serve as a canvas upon which the mind reconciles its experiences, reflects on its challenges, and engages in a form of self-therapy—a window into the depths of the human psyche, inviting us to explore the inner landscapes that shape our waking lives.

PSYCHEDELICS AND EXPANDED AWARENESS:

Historically, the use of psychedelic substances to induce altered states of consciousness goes as far back as thousands of years. In this part of the book, we investigate the resurgence of interest in psychedelics for therapeutic purposes, as well as their potential to catalyze profound shifts in perception and self-awareness.

PSYCHEDELICS: Unveiling the Resurgence and Profound Potential for Perception and Healing

For millennia, humanity has engaged with psychedelic substances to explore altered states of consciousness, seeking insight, spiritual connection, and expanded perception. The resurgence of interest in psychedelics, particularly for therapeutic purposes, represents a remarkable shift in understanding their potential to catalyze profound shifts in perception, self-awareness, and healing. Let's now discover the exploration of psychedelics' resurgence and their transformative possibilities. As ongoing research navigates the complexities of their use, we are confronted with a realm of transformative possibilities—one that invites us to explore the depths of our minds and the mysteries of human consciousness.

The human mind, with its mysteries and complexities, remains a frontier of exploration, inviting us to journey inward and unravel the profound enigmas that define our existence. As we venture into the depths of consciousness, traverse the neural landscapes of the brain, and contemplate altered states of awareness, we gain new insights into the nature of our minds and the intricate tapestry of human cognition. We invite you to come with us as we unlock the doors to the mind's labyrinth, where each step brings us closer to understanding the essence of who we are and the wonders that lie within.

RESURGENCE OF INTEREST:

SCIENTIFIC EXPLORATION: The mid-20th century witnessed extensive scientific research into psychedelics like LSD and psilocybin. However, due to social and political factors, this research was halted, only to experience a revival in recent years as regulatory barriers eased and researchers rekindled their interest.

THERAPEUTIC PROMISE: Studies have shown that psychedelic-assisted therapy can be effective in treating conditions like depression, anxiety, PTSD, and addiction. Research indicates that these substances, when administered in controlled settings, may facilitate breakthroughs in psychotherapy by inducing states of introspection and emotional release.

PROFOUND SHIFTS IN PERCEPTION AND SELF-AWARENESS:

ALTERED STATES OF CONSCIOUSNESS: Psychedelics induce altered states of consciousness that can lead to profound perceptual changes. Users often describe heightened sensory experiences, synesthesia (cross-sensory perceptions), and a dissolution of boundaries between self and the external world.

EGO DISSOLUTION: Some psychedelics, particularly substances like psilocybin and DMT, can induce a sense of ego dissolution—an experience of transcending one's individual identity. This can lead to a deep sense of interconnectedness and unity with the universe, fostering a profound shift in self-awareness.

MYSTICAL AND SPIRITUAL EXPERIENCES: Many users report having mystical or spiritual experiences while under the influence of psychedelics. These experiences, characterized by a sense of awe, transcendence, and encounters with a divine presence, can have lasting impacts on an individual's worldview and values.

NEUROSCIENTIFIC INSIGHTS:

DEFAULT MODE NETWORK (DMN): Psychedelics are known to modulate the DMN—the brain network associated with self-referential thinking and mind-wandering. Disruption of the DMN during a psychedelic experience might explain the altered sense of self and interconnectedness reported by users.

NEUROPLASTICITY AND CONNECTIVITY: Psychedelics seem to enhance neuroplasticity—the brain's ability to rewire itself. They may create temporary states of heightened neural plasticity, allowing for new connections and insights to form.

CHALLENGES AND ETHICAL CONSIDERATIONS:

SAFETY AND REGULATION: While research indicates therapeutic potential, psychedelics are potent substances with potential risks. Ensuring safe administration, proper set and setting, and careful screening of participants are critical for their responsible use.

Cultural Context: The cultural, historical, and spiritual significance of psychedelics raises questions about cultural appropriation, respect for indigenous traditions, and ethical considerations in their use.

A Catalyst for Healing and Transformation:

Therapeutic Potential: Psychedelic-assisted therapy offers a unique opportunity to explore and process deep-seated emotions, trauma, and psychological challenges in a controlled and supportive environment.

Spiritual and Personal Growth: Beyond therapy, psychedelic experiences can lead to profound shifts in personal values, beliefs, and perspectives, often fostering a sense of interconnectedness and purpose.

Modern Psychedelic Research: Unveiling the Legacy of Albert Hofmann's Discovery

The modern era of psychedelic research traces its roots back to a pivotal moment in 1938 when Swiss chemist Albert Hofmann synthesized lysergic acid diethylamide (LSD-25) for the first time. This discovery marked the beginning of a journey that would ultimately lead to a renaissance in understanding the potential of psychedelic substances for therapeutic, psychological, and spiritual exploration. Let's embark on an exploration of this remarkable journey.

ALBERT HOFMANN'S DISCOVERY OF LSD:

ACCIDENTAL REVELATION: Albert Hofmann initially synthesized LSD-25 as part of his research into ergot alkaloids, substances derived from a fungus that grows on rye. His accidental exposure to a minute amount of LSD, leading to a unique altered state of consciousness, marked the first recorded intentional LSD trip.

LSD'S EARLY USES: In the years following his discovery, LSD found applications in psychiatric research and psychotherapy. Researchers believed that LSD might offer insights into mental illness and provide therapeutic benefits.

INITIAL WAVE OF RESEARCH:

PSYCHIATRIC AND THERAPEUTIC EXPLORATION: In the 1950s and 1960s, researchers explored the therapeutic potential of LSD for various conditions, including anxiety, depression, and alcohol addiction. Some studies reported positive outcomes, suggesting that LSD-assisted psychotherapy could be transformative.

PSYCHEDELICS IN PSYCHOLOGY AND SPIRITUALITY: Psychedelics, including LSD, were also embraced by psychologists and spiritual seekers. Psychedelic experiences were seen as a means to access altered states of consciousness, inner exploration, and potential spiritual insights.

Shifts in Perception and Regulation:

Counterculture Movement: The 1960s saw a surge in interest in psychedelics, fueled by the counter-culture movement. Psychedelics became associated with anti-establishment sentiments, artistic expression, and alternative spirituality.

Regulatory Restrictions: Due to concerns over misuse, the potential for adverse effects, and the lack of standardized research protocols, regulatory restrictions were put in place, and regulatory authorities began to restrict the use of psychedelics. As a result, Research faced increasing challenges and limitations.

Revival of Research (Recent resurgence): Starting in the late 20th century and accelerating in the 21st century is a resurgence of interest in psychedelics for therapeutic and research purposes emerged. Advances in neuroscience, new research methodologies, and a reevaluation of the therapeutic potential of psychedelics contributed to this revival.

Modern research has explored the efficacy of psychedelic-assisted therapy for conditions like depression, PTSD, anxiety, and addiction. Studies suggest that under controlled settings, psychedelics may facilitate breakthroughs and healing.

Psychological and neuroscientific Insights have highlighted the relationship between neuroplasticity and brain connectivity. Psychedelics have therefore shown to modulate brain networks, such as the default mode network, potentially contributing to shifts in perception and self-awareness.

Ego Dissolution and Mystical Experiences: Research indicates that psychedelics can induce ego dissolution—an experience of transcending individual identity—and mystical or spiritual experiences that can lead to lasting-psychological shifts.

Ethical Considerations and Future Directions:

Modern psychedelic research emphasizes the importance of ethical considerations, including participant safety, informed consent, and respecting cultural and indigenous traditions.

Recent research has rekindled hope for the therapeutic potential of psychedelics, leading to groundbreaking studies, pilot programs, and even regulatory approvals for treatments like psilocybin-assisted therapy.

Albert Hofmann's accidental discovery of LSD set in motion, a journey of scientific exploration, social upheaval, regulatory challenges, and, ultimately, a renaissance of modern psychedelic research. As contemporary studies delve deeper into the therapeutic, psychological, and spiritual implications of psychedelic substances, we stand at pivotal juncture where science and culture converge to unveil the profound potential of these substances to enhance our understanding of consciousness, healing, and the human experience.

Chapter 8

Preserving Our Planet for the Future

Generations

Nurturing Sustainability, Combating Climate Change, and Fostering Global Responsibility

In a world confronted by pressing environmental challenges, the imperative to preserve our planet for future generations has never been more crucial. This chapter delves into the interconnected issues of sustainability, climate change, renewable energy, and the role that both global cooperation and individual actions play in safeguarding our planet's delicate balance.

ENVIRONMENTAL CHALLENGES AND THE IMPERATIVE OF SUSTAINABILITY

Ecosystem Degradation: The degradation of ecosystems, from deforestation to habitat loss, threatens biodiversity and disrupts delicate ecological balances essential for planetary health.

The delicate web of life on Earth depends on the intricate interactions within ecosystems, which encompasses everything from forests and oceans to grasslands and wetlands. However, ecosystem degradation, marked by processes like deforestation and habitat loss pose a grave threat to biodiversity and disrupts the delicate ecological balances that are essential for planetary health. Let's delve into the complex interplay of ecosystem degradation, its consequences, and potential remedies.

Deforestation and Habitat Loss

The widespread clearing of forests for agriculture, logging, urbanization, and infrastructure development lead to the loss of critical habitats. Trees, which provide homes for countless species and contribute to carbon sequestration, are lost at an alarming rate.

Habitat Fragmentation is another threat, and this occurs when habitats are divided into smaller, isolated patches due to human activities. This disrupts natural corridors for species migration and reduces the availability of resources, leading to isolation and population decline.

CONSEQUENCES FOR BIODIVERSITY:

SPECIES EXTINCTION: Habitat loss is a primary driver of species extinction. Many species, especially those with specific habitat requirements, struggle to adapt to altered landscapes and face heightened risks of disappearing.

Keystone species, which play disproportionately important roles in maintaining ecosystem structure, can suffer from habitat loss. Their decline ripples through entire ecosystems, impacting other species and ecosystem functions.

DISRUPTION OF ECOLOGICAL BALANCES:

Ecosystems rely on complex food webs, where species interactions maintain balance. Habitat loss can disrupt these interactions, leading to the overpopulation of some species and underpopulation of others, upsetting the equilibrium.

Ecosystems provide a range of services—pollination, water purification, climate regulation—that underpin human well-being. Disruption of these services due to habitat loss can have far-reaching consequences.

REMEDIES FOR ECOSYSTEM DEGRADATION:

Establishing and effectively managing protected areas, such as national parks and wildlife reserves, can safeguard critical habitats and provide safe havens for species.

Habitat restoration efforts involve replanting native vegetation, creating wildlife corridors, and rehabilitating degraded areas to promote habitat connectivity and restore ecosystem functions.

Promoting sustainable agricultural and forestry practices ensure that human activities coexist with ecosystem preservation, mitigating the impact of habitat loss.

CONSERVATION POLICIES: Government policies that enforce responsible land use, regulate logging and development, and incentivize sustainable practices are vital in combatting habitat loss.

Engaging local communities in conservation efforts fosters a sense of ownership and responsibility, leading to more effective protection of ecosystems.

Ecosystem degradation knows no borders and therefore international cooperation is essential to combat issues like deforestation, hence countries must work together to halt illegal logging and protect transboundary habitats.

RAISING AWARENESS: Education and public awareness campaigns play a pivotal role in garnering support for conservation efforts, urging individuals to make sustainable choices, and advocating for policy change.

A CALL TO RESTORE BALANCE

As we witness the consequences of ecosystem degradation,

it becomes imperative to act swiftly and decisively. By addressing habitat loss, promoting conservation, and fostering sustainable practices, we can restore the delicate ecological balances that sustain life on Earth. Every effort to protect ecosystems, no matter how small, contributes to the intricate web of life and the preservation of planetary health—a shared responsibility for the well-being of current and future generations.

Overexploitation of finite exacerbates environmental challenges posing long- term threats to economic stability.

The overexploitation of finite resources, including freshwater, minerals, and fossil fuels, has become a pressing concern with far-reaching implications for both the environment and global economies. This chapter looks into the intricate web of challenges stemming from overexploitation, its environmental and economic consequences, and the potential solutions that can help us navigate these complex issues.

WATER RESOURCES DEPLETION:

As populations increase and industrialization expands, demand for freshwater surges.Agricultural irrigation, industrial processes, and domestic consumption strain available water sources.

Over-extraction of groundwater from aquifers, often faster than they can recharge, leads to sinking land, compromised water quality, and depletion of a vital resource.

Mineral Resource Depletion adds another layer to the problem.

Mining for minerals fuels industries from electronics to construction. However, many minerals are finite, and their extraction often leads to habitat destruction, soil erosion, and toxic waste.

The demand for certain minerals, like tantalum and coltan, has driven conflicts in resource- rich regions, exacerbating social and geopolitical instability.

Fossil Fuel Depletion and Climate Impact:

Fossil fuels, including oil, coal, and natural gas, are finite resources. Overreliance on these non-renewable sources has led to geopolitical conflicts and environmental degradation.

The combustion of fossil fuels releases greenhouse gases, driving climate change and exacerbating environmental challenges like sea-level rise and extreme weather events.

Consequences for Economic Stability:

Overreliance on finite resources can lead to price volatility as scarcity increases. Fluctuating prices impact industries, trade, and consumer markets.

Economies heavily reliant on resource extraction face vulnerability to market fluctuations and the risk of "resource curse"—economic instability despite resource wealth.

SOLUTIONS TO ADDRESS RESOURCE DEPLETION:

Efficient use of resources, from water to minerals, is paramount. Technologies that reduce waste and optimize consumption can extend the life-span of these resources.

TRANSITION TO RENEWABLE ENERGY: There has been a recent shift from fossil fuels to renewable energy sources like solar, wind, and hydropower. This is a step in the right direction, but a lot still needs to be done in these domains to alleviate dependence on finite resources while curbing emissions.

Implementing circular economy principles (e.g., recycling, reusing, and reducing waste) minimizes the demand for new resources and reduces devastating environmental impact.

Mining industries can adopt sustainable practices, minimizing habitat destruction and prioritizing reclamation and restoration.

Efficient irrigation systems and crop selection can reduce water consumption in agriculture, a sector that often contributes significantly to water depletion.

Global Collaboration: International Agreements and frameworks can help regulate resource extraction, minimize conflicts, and promote responsible resource management.

INNOVATION AND TECHNOLOGY: Advancements in technology, from desalination for freshwater to improved extraction methods, can enhance resource availability and efficiency.

Resource depletion is a complex challenge that requires concerted efforts at local, national, and global levels to combat it. By embracing sustainable practices, transitioning to renewable energy, and fostering responsible management, we can alleviate the strain on finite resources, mitigate environmental impacts, and ensure long-term economic stability. As custodians of the planet, we hold the power to reshape our relationship with resources, and to forge a more harmonious and sustainable future for all.

POLLUTION AND WASTE:

Pollution of air, water, and soil, coupled with the alarming accumulation of waste has become a pressing global challenge that imperils ecosystems and human health. In this part of the book, we take a look at the multifaceted consequences of environmental pollution and waste, highlighting the intricate interplay of these issues and the potential solutions that can help restore our planet's health.

AIR POLLUTION AND ITS RAMIFICATIONS:

Sources of air pollution for instance industries, vehicles, and the burning of fossil fuels release pollutants like particulate matter, nitrogen oxides, and volatile organic compounds into the air.

The impact of air pollution on human health is huge. Air pollution leads to respiratory illnesses, cardiovascular diseases, and even premature death, and vulnerable populations like children and the elderly at higher risk.

WATER POLLUTION AND ITS RIPPLE EFFECTS:

Sources of water contaminants may include but are not limited to industrial runoffs agricultural pesticides, and untreated sewage discharge pollute water bodies, compromising their quality and harming aquatic life.

HUMAN HEALTH IMPACT: Polluted water sources can cause waterborne diseases, affecting millions worldwide and undermining communities' access to clean water.

SOIL POLLUTION AND ECOLOGICAL DISRUPTION:

INDUSTRIAL CHEMICALS: Pesticides, heavy metals, and hazardous waste infiltrate soil, posing threats to soil quality and plant growth.

BIODIVERSITY LOSS: Soil pollution disrupts the delicate balance of ecosystems, impacting soil-dwelling organisms and the health of plants that rely on healthy soils.

WASTE ACCUMULATION AND ITS TOLL:

MOUNTAINS OF WASTE: The accumulation of nonbiodegradable waste, especially plastic, threatens both terrestrial and marine ecosystems, harming wildlife and marine life.

MICROPLASTICS AND TOXICITY: Plastic waste breaks down into microplastics that infiltrate food chains and pose risks to both animal and human health.

SOLUTIONS TO ADDRESS POLLUTION AND WASTE:

REDUCING EMISSIONS: Enforcing stricter emission standards for industries and vehicles helps curb air pollution. Transitioning to renewable energy sources also mitigates pollution.

WASTEWATER TREATMENT: Treating industrial and domestic wastewater before discharge into water bodies minimizes water pollution and protects aquatic ecosystems.

REGULATING CHEMICAL USE: Stringent regulations on chemical use in agriculture and industry can prevent soil pollution and protect both ecosystems and human health.

WASTE MANAGEMENT: Adopting comprehensive waste management strategies, including recycling, composting, and responsible disposal, reduces waste accumulation.

PLASTIC REDUCTION: Initiatives to reduce single-use plastics and promote alternatives contribute to curbing plastic waste and its ecological impact.

Awareness and Education: Raising public awareness about the impact of pollution and waste fosters responsible consumption and drives policy change.

Policy Enforcement: Governments must enforce environmental regulations and impose penalties for violations to deter pollution and improper waste management.

A Shared Commitment:

Addressing environmental pollution and waste requires a collective commitment to change. By embracing sustainable practices, advocating for policy reforms, and fostering a deeper sense of responsibility for the well-being of our planet, we can reverse the damage caused by pollution and waste. Through individual and collective actions, we can create a world where ecosystems thrive, human health flourishes, and the legacy we leave behind is one of stewardship and restoration.

Climate Change, Renewable Energy, and Global Cooperation

The Climate Crisis: Unveiling the Existential Threat and Pathways to Mitigation

Climate change, driven by human activities such as burning fossil fuels, deforestation, and industrial emissions, presents an existential threat to our planet with consequences that reach far beyond environmental changes. This chapter continues to look into the intricate web of challenges posed by climate change, highlighting potential existential threats and offering pathways to avert them.

EXISTENTIAL THREATS OF CLIMATE CHANGE:

RISING TEMPERATURES: Global warming intensifies heatwaves, droughts, and wildfires, endangering ecosystems, agriculture, and human health.

SEA-LEVEL RISE: Melting glaciers and ice sheets cause sea levels to rise, submerging coastlines, displacing communities, and threatening cities.

EXTREME WEATHER: More frequent and severe storms, hurricanes, and floods disrupt communities, infrastructure, and economies.

BIODIVERSITY LOSS: Altered habitats, changing temperatures, and ocean acidification drive species extinction, affecting ecosystems' resilience and services.

FOOD AND WATER INSECURITY: Climate-related impacts on agriculture and freshwater resources lead to food scarcity and conflicts over water access.

Pathways to Avert Existential Threats:

TRANSITION TO RENEWABLE ENERGY:

RENEWABLE REVOLUTION: Embracing renewable energy

sources like solar, wind, and hydroelectric power is pivotal for curbing greenhouse gas emissions (reducing carbon footprint) and transitioning to a sustainable energy future.

POLICY SUPPORT: Governments can incentivize renewable energy adoption through subsidies and regulations. Governments play a pivotal role in enacting policies that incentivize sustainable practices, curb emissions, and hold industries accountable for their ecological impact

CARBON CAPTURE AND STORAGE (CCS):

CAPTURINGEMISSIONS: CCS technologies trap CO2 emissions from industrial processes and power plants.

LONG-TERM STORAGE: Captured CO2 can be stored underground to prevent its release into the atmosphere.

REFORESTATION AND FOREST CONSERVATION:

RESTORING ECOSYSTEMS: Reforestation and afforestation efforts sequester carbon and restore habitats.

HALTING DEFORESTATION: Protecting existing forests prevents carbon release and safeguards biodiversity.

SUSTAINABLE AGRICULTURE:

CLIMATE-SMART FARMING: Implementing practices that reduce emissions, enhance soil health, and optimize resource use.

DIVERSE CROPS: Promoting diverse crop varieties increases resilience to changing climate conditions.

ADAPTING TO CHANGES:

CLIMATE-RESILIENT INFRASTRUCTURE: Designing infrastructure to withstand climate-related hazards.

COMMUNITY PLANNING: Integrating climate risks into urban and rural planning to minimize vulnerability.

INTERNATIONAL COLLABORATION:

PARIS AGREEMENT: Global commitment to limit temperature rise by reducing emissions.

CLIMATE DIPLOMACY: International cooperation is vital to address transboundary impacts and share technological solutions.

CONSUMER CHOICES AND LIFESTYLE CHANGES: Individual Actions as Catalysts for Change

REDUCED CONSUMPTION: Choosing sustainable products, minimizing waste, and embracing energy-efficient practices. Individuals wield significant power through their choices as consumers. Opting for sustainable products, reducing single-use plastics, and supporting eco-friendly practices can drive positive change.

ADVOCACY AND EDUCATION: Raising awareness about the

impact of individual actions on climate change. Raising awareness about environmental issues and advocating for policies that prioritize sustainability is vital. Education empowers individuals to influence both policymakers and industries.

URGENT AND COLLECTIVE ACTION:

Addressing the existential threat of climate change requires swift, comprehensive, and collaborative efforts at global, national, and local levels. By shifting toward sustainable practices, embracing clean energy solutions, and advocating for policy changes, we can avert the most severe consequences of climate change. As stewards of our planet, we hold the power to forge a future where ecosystems flourish, communities thrive, and the legacy we leave behind is one of responsible environmental stewardship.

GLOBAL COLLABORATION: Addressing climate change necessitates international cooperation. Agreements like the Paris Agreement highlight the urgency of a united front in mitigating climate impacts.

COMMUNITY ENGAGEMENT: Grassroots movements and community initiatives contribute to environmental change. Community gardens, clean-up drives, and conservation efforts showcase the strength of collective action.

BALANCING PROGRESS WITH PLANETARY STEWARDSHIP:

CIRCULAR ECONOMY: Shifting from a linear "take-make-dispose" economy to a circular one promotes recycling, reusing, and reducing waste, minimizing the strain on natural resources.

GREEN TECHNOLOGIES: Innovations in sustainable technologies, from electric vehicles to green building practices, reduce environmental footprints and drive economic growth.

GUARDIANS OF THE FUTURE:

INTERCONNECTEDNESS: Understanding humanity's interdependence with the natural world reinforces the importance of sustainable practices for future generations.

LEGACY OF STEWARDSHIP: As custodians of our planet, we hold the responsibility to safeguard its wonders and resources for those who will follow us.

HOPE AND ACTION: This chapter concludes with a call to action, urging readers to engage in meaningful endeavors that preserve the Earth's vitality and diversity for the well-being of generations to come.

In the pursuit of a sustainable future, understanding the gravity of environmental challenges, harnessing renewable energy, fostering global cooperation, and embracing individual actions are all vital pieces of the puzzle. By uniting our efforts to cultivate responsible habits, advocate for change, and champion ecological preservation, we can leave a legacy of stewardship that ensures the beauty and vitality of our planet endure for generations yet to unfold.

219

Embracing the Legacy of Knowledge and Wonder

As we reach the culmination of our journey through time, science, and wonder, we find ourselves at the crossroads of reflection and anticipation. This journey has been a quest to unravel the secrets of humanity, to venture beyond the boundaries of what we know, and to stand in awe of the vast expanse of knowledge that surrounds us.

REFLECTING ON OUR JOURNEY:

RECAPPING THE WONDERS: Our journey has taken us from the origins of humanity, through the scientific revolution, the mysteries of the universe, the marvels of biology and medicine, the power of technology, the enigma of the human mind, and the imperative of environmental stewardship. Each chapter unveiled a tapestry of discoveries that have shaped our understanding of who we are and our place in the cosmos.

UNDERSTANDING FOR A BETTER FUTURE: Through exploring the annals of history and the insights of science, we've glimpsed the interconnectedness of all life and the responsibility we bear for our planet. Understanding our past and present empowers us to forge a more sustainable and compassionate future, where the lessons of history guide our choices and the potential of science fuels our progress.

EMBRACING THE JOURNEY AHEAD:

CONTINUED PURSUIT OF KNOWLEDGE: As this book concludes, let it not mark the end of our quest for understanding but rather the spark that ignites a lifelong pursuit of knowledge. The universe remains a boundless playground of mysteries waiting to be explored, and every discovery deepens our connection to the tapestry of existence.

CULTIVATING WONDER AND AWE: The journey we embarked upon reminds us of the profound wonder that saturates our world. From the intricate biology within us to the majesty of the cosmos above, embracing the mysteries that remain evokes a sense of awe that enriches our lives and inspires our endeavors.

In the intersection of knowledge and curiosity, of science and wonder, lies the essence of what it means to be human. The pursuit of understanding propels us forward, yet it is our capacity for awe that gives depth to our existence. As we bid farewell to these pages, let us carry with us the torch of curiosity, the flame of discovery, and the spirit of wonder that has guided us through this journey.

For in our ceaseless quest to unravel the secrets of

humanity and the universe, we not only seek knowledge but also cultivate a profound appreciation for the beauty, complexity, and interconnectedness of all things. May this journey be a reminder that, in the tapestry of existence? We are both explorers and custodians, seekers of wisdom and stewards of wonder.

Epilogue

A Letter to Future Generations

Dear Future Explorers of the Cosmos,

As we conclude this journey through time, science, and wonder, we stand at a pivotal juncture where the threads of our understanding meet the canvas of what lies ahead. This book, "Unraveling the Secrets of Humanity," has been a testament to the curiosity that unites us across generations, the pursuit of knowledge that propels us forward, and the boundless wonder that colors our existence.

1. Addressing the Legacy:

As we pen these words, we acknowledge that the legacy we leave for you is a tapestry woven from the threads of our collective quest for understanding. In the chapters of this book, we've shared the stories of our past, the breakthroughs that shaped our present, and the challenges that underscore the importance of responsible stewardship.

Every decision, every discovery, and every commitment to knowledge today echoes through the corridors of time, resonating with the intent to preserve, protect, and propel.

2. Encouraging Ongoing Exploration:

Our journey doesn't end with these pages. Instead, it becomes a steppingstone into a future where knowledge remains a beacon, illuminating the path of discovery. We encourage you to carry forward the torch of inquiry, to gaze into the night sky with the same wonder that inspired us, and to delve deeper into the mysteries of existence. Embrace new technologies, challenge established paradigms, and venture into realms yet unexplored. Let the timeless spirit of exploration guide you as it has guided us.

3. Hope for a Future Full of Promise:

In the unfurling tapestry of the universe, each generation weaves its thread, contributing to a narrative of progress and potential. With the dawn of each new day, you have the opportunity to shape a future full of promise, ingenuity, and compassion. Our fervent hope is that you will inherit a world where the delicate balance between humanity and nature is restored, where scientific advancements continue to illuminate the unknown, and where the wonders of existence inspire both awe and responsible action.

As you read these words, remember that you stand on the shoulders of countless explorers who came before you— those who dared to question, those who dared to dream, and those who dared to make a difference. Let their courage embolden you as you navigate the complexities of your time.

With each question you ask, each experiment you conduct, and each boundary you push, you contribute to an ever-evolving story that began with the first stirrings of curiosity in human hearts. The pursuit of knowledge and the embrace of wonder will forever be our legacy to you—a legacy of hope, of discovery, and an unquenchable thirst for understanding.

May your journey be as awe-inspiring and transformative as the one we have shared within these pages. For as long as you seek, explore, and ponder, the spirit of this journey will thrive—illuminating the path toward a future that is truly worth unraveling.

With boundless hope and unwavering admiration,

Felix Atoh